Irreplaceable

How AI Changes Everything (and Nothing) in Teaching and Learning

Maya Bialik
Peter Nilsson

Solution Tree | Press
a division of Solution Tree

555 North Morton Street
Bloomington, IN 47404
800.733.6786 (toll free) / 812.336.7700
FAX: 812.336.7790
email: info@SolutionTree.com
SolutionTree.com
Printed in the United States of America

Library of Congress Cataloging-in-Publication Data

Names: Bialik, Maya author | Nilsson, Peter (Educator) author
Title: Irreplaceable : how AI changes everything (and nothing) in teaching and learning / Maya Bialik, Peter Nilsson.
Description: Bloomington, IN : Solution Tree Press, 2026. | Includes bibliographical references and index.
Identifiers: LCCN 2025029167 (print) | LCCN 2025029168 (ebook) | ISBN 9798893740394 paperback | ISBN 9798893740400 ebook
Subjects: LCSH: Artificial intelligence--Educational applications
Classification: LCC LB1028.44 .B53 2026 (print) | LCC LB1028.44 (ebook) | DDC 371.33/4--dc23/eng/20251215
LC record available at https://lccn.loc.gov/2025029167
LC ebook record available at https://lccn.loc.gov/2025029168

Solution Tree
Cameron L. Rains, CEO
Edmund M. Ackerman, President
Solution Tree Press
Publisher: Kendra Slayton
Associate Publisher: Todd Brakke
Acquisitions Director: Hilary Goff
Editorial Director: Laurel Hecker
Art Director: Rian Anderson
Managing Editor: Sarah Ludwig
Copy Chief: Jessi Finn
Production Editor: Madonna Evans
Copy Editor: Anne Marie Watkins
Proofreader: Elijah Oates
Cover and Text Designer: Abigail Bowen
Content Development Specialist: Amy Rubenstein
Associate Editor: Elijah Oates
Editorial Assistant: Madison Chartier

Acknowledgments

We want to express our gratitude to educators everywhere. Education is the most important work, and it is also among the greatest joys. We are proud to have dedicated our professional lives to serving students and teachers, and we are grateful for our mentors and colleagues.

We would also like to thank the researchers and developers advancing the field of artificial intelligence. It has been a thrill to research, explore, experiment, learn, create, and so much more with the support of such a multifaceted partner. Equally, we'd like to thank the researchers studying teaching and learning, whose systematic inquiry has yielded both validating and surprising insights that have added layers of richness to the conversation.

This book also wouldn't be what it is without feedback during the writing process from Maha Bali, Justin Cerenzia, Nat Damon, Andy Housiaux, Eric Hudson, Kristina Ishmael, Justin Reich, Indu Singh, Darri Stephens, Christian Talbot, and others. If there is wisdom in this book, it has benefited from their contributions. If there are errors or shortcomings in this book, they are ours.

Our thanks also to Solution Tree: Hilary Goff, Madonna Evans, Abby Bowen, Rian Anderson, and the rest of the team. Our gratitude for your support and encouragement.

We would like to thank Charles Fadel for introducing us and for his work leading the Center for Curriculum Redesign, which both of us have been grateful to be a part of in different capacities.

Finally, we would like to thank our spouses for putting up with long writing hours—especially when deadlines were approaching.

Solution Tree Press would like to thank the following reviewers:

Zachary Ashauer
Social Sciences Teacher
Hortonville Area School District
Hortonville, Wisconsin

Amber Gareri
K–12 Science Coordinator
Pasadena Independent School District
Pasadena, Texas

Janet Gilbert
Lead Instructional Coach
Deer Valley Unified School District
Phoenix, Arizona

Kelly Hilliard
Gifted and Talented Mathematics Instructor
Darrel C. Swope Middle School
Reno, Nevada

Brad McDiarmid
Professor, School of Arts and Education
Red Deer Polytechnic
Red Deer, Alberta, Canada

Nicole McRee
Science, STEM, and Wellness
Curriculum Specialist
Kildeer Countryside Community
Consolidated School District 96
Buffalo Grove, Illinois

Janet Nuzzie
Intervention Specialist, K–12 Mathematics
Pasadena Independent School District
Pasadena, Texas

Allison Zamarripa
Reading and Language Arts Specialist
Pasadena Independent School District
Pasadena, Texas

Table of Contents

About the Authors

Maya Bialik is the founder of QuestionWell, which creates research-aligned AI tools for education that improve quality, increase agency, and strengthen human connection. A former middle school science teacher, Maya is now completing her PhD with a focus on how AI can improve working conditions for teachers, making their creative work easier, more effective, and more enjoyable. Her two previous books are *Four-Dimensional Education: The Competencies Learners Need to Succeed* (2015) and *Artificial Intelligence in Education: Promises and Implications for Teaching and Learning* (2019).

Since earning her master's degree in mind, brain, and education from Harvard Graduate School of Education in 2012, Maya has been passionate about bridging the gaps among researchers, teachers, school leaders, policymakers, and technologists. Combining her immigrant perspective and her experience learning, performing, and directing narrative and musical improvisation, she creates collaborative spaces for grappling with complex topics and building mutual understanding.

Maya serves on the board of The Plenary, Co., a civic arts and sciences nonprofit building collective knowledge, community agency, and public imagination. She is also an adviser for the Center for Curriculum Redesign, a nonprofit that asks, "What should students learn to be prepared for the age of AI?" She holds bachelor's degrees in psychology, biology, and linguistics from the University of Massachusetts Amherst.

Peter Nilsson is an educator, writer, and musician with over two decades of experience in teaching and school leadership. He was the head of school at King's Academy, a coeducational boarding school in Jordan, and he taught English for fifteen years and served in leadership positions at Deerfield Academy in Massachusetts. Peter is also the editor of *The Educator's Notebook*, a weekly newsletter that gathers education- and learning-related news from around the web for the purpose of promoting innovation in education. He is the cofounder of Athena Lab, a collaborative online platform for teachers.

Peter serves on the advisory boards for SXSW EDU, the Center for Curriculum Redesign, and the Middle States Association's Responsible AI in Learning endorsement. His presentations on AI in education, creative work, cognitive science, and educational technology have been featured at conferences in the United States and abroad, including SXSW EDU, the College Board, and the Organisation for Economic Co-operation and Development (OECD). In 2017, he became the first high school teacher to receive the Wolfram Innovator Award.

A composer and writer, Peter is also the creator of a musical about schools. His creative work in music and education reflects his belief that the most powerful learning happens at the intersections of disciplines, cultures, and ideas.

Peter studied English and music at Middlebury College, earned his master of arts from Columbia University, and received a National Endowment for the Humanities institute grant to study *Paradise Lost.*

To book Maya Bialik or Peter Nilsson for professional development, contact pd@SolutionTree.com.

Introduction

If it's out of the bag, then it's out of the bag.

—AESOP ROCK

This is a book about teachers and teaching, for if the body were a school, teachers would be the beating heart through which students circulate, year in and year out. This is also a book about students and learning. More specifically, it is about how the emerging technology of our time, artificial intelligence (AI), changes the experience of teachers teaching and students learning, as well as the interactions among them.

When we tell people about this book, many ask, "Do you think AI will replace teachers?" It's a provocative question that reveals how people think about teaching, learning, and the purposes of schools.

To answer directly—no. AI will not replace teachers. We argue that the core components of teaching and learning—the agency of students and teachers and the relationships among them—are *irreplaceable*. So much so that we gave our book that title.

But that's not the whole story. AI will replace tasks, and it has the potential to substantially change the experience of school—in many ways for the better and in many other ways for the worse. Our aim is to help bring about the most positive alternative future (Facer, 2011). This future won't arrive on its own; we must purposefully imagine it (Jasanoff & Kim, 2015). In the words of Reid Hoffman and Greg Beato, "You'll never get the future you want simply by prohibiting the future you *don't* want" (2025, p. xiii). So, without dismissing the potential risks, this book identifies possible benefits of AI in teaching and learning and charts a path toward them.

To reach a positive alternative future, we offer insights drawn from our experiences in the classroom and in the technology sector, as well as from academic research about learning, the brain, and technology. As a result, this book is not only about the art of teaching or the science of learning but the ways AI interacts with both teaching and learning. By exploring this intersection, we aim to propel discourse in schools toward a deep understanding of what AI-empowered teaching and learning can (and, in some cases, already does) look like.

The More Things Change . . .

Technologies have transformed the learning experience before. *Transformed* may be an overused word in the 21st century world of overlapping hype cycles, but a look back over history shows that at pivotal moments, it is true (Dedehayir & Steinert, 2016).

- Writing shifted how we manage information, reshaping our relationship with the brain's capacity for memory.
- The printing press democratized information, making more diverse texts more accessible and changing what ideas became available to everyone.
- The dictionary and encyclopedia aggregated and organized knowledge, making its contours conceivable and navigable.
- The calculator commodified computational ability, bringing both beginner and advanced mathematics within reach of more people.
- The word processor and personal computer accelerated the process of writing, enabling easier and more efficient iteration of complex thought.
- The internet enabled greater information sharing and asynchronous and remote collaboration (Gleick, 2011; Ong, 1982).

What does it mean to transform learning? Consider the invention, the *technology*, of writing, which emerged around five thousand years ago—just a breath in the song of human life on earth. The popularization of writing spawned great debates about the impact it would have. Socrates notably refused to record his thoughts because he believed writing would weaken memory and lead readers to misinterpret meaning, since the author could not clarify and tailor their message to their audience (Plato, ca. 370 BCE/2002). He was right. People from cultures with a writing system have much more limited memory capabilities than those from oral cultures (Ong, 1982), and we are constantly misunderstanding each other's writings. But thankfully, Plato wrote down what Socrates said, because without Plato having recorded Socrates's thoughts, we would not have been able to learn from Socrates thousands of years later. Writing reduces our use of memory and opens our words up to misinterpretation, but it has also enabled the preservation of thought over time, allowing the accumulation of knowledge, the development of science, the diffusion of religion, and the circulation of literature. Writing changed what, how, and even why we learn (Ong, 1982).

Now, in the early 21st century, AI could have the most significant impact on learning since the invention of writing.

Our goal with this book is to help prepare teachers for this transformative technology. Written for teachers, school leaders, and those who work with teachers and school leaders, this book provides our best insights for understanding and enhancing teacher work, student work, and class time in a time of AI. We also recognize that technology changes rapidly, so we introduce practical approaches to AI in the context of principled frameworks that we hope are as applicable in twenty years as they are over the next two years.

. . . The More They Stay the Same

Despite the dramatic technological changes over the decades and centuries, the skills we teach students have remained largely the same. Critical thinking, communication, and collaboration are timeless. Whether students work with parchment or computers, we, as teachers, have asked them to analyze evidence, creatively solve problems, evaluate sources, engage with complex ideas, and express themselves accurately and powerfully (Fadel, Bialik, & Trilling, 2015). Now, our modern curricula have deepened our understanding of what is timeless, including social-emotional learning, higher-order thinking skills, the development of dispositions and character qualities, and the awareness and practice of metacognition and meta-learning (Fadel et al., 2015).

None of this changes with new technologies; what changes are the tools we use. Before the graphing calculator, we asked students to draw graphs and then interpret them. Now, students use graphing calculators to draw graphs, and then they interpret them. Before the internet, we taught students to find sources in the library and analyze them. Now, with the internet, we teach students to find sources at the library *and* online and then analyze them. Before generative AI, students wrote essays through iterative drafting processes. Now, with generative AI, they still write essays through iterative drafting processes—just with AI as a support. Our objectives remain the same no matter how rapidly the technological world advances.

These changing technologies allow us to do new and different kinds of work, but they do not change the core mechanisms that drive meaningful learning and shape our human experience. Learning happens through actions, behaviors, and habits. It is slow. Technology can make learning more engaging (Schindler, Burkholder, Morad, & Marsh, 2017), optimize study strategies and schedules (Ma, Adesope, Nesbit & Liu, 2014; Tabibian et al., 2019), and even strategically activate, strengthen, and weaken systems in the brain (Lembke, 2021)—but it cannot make learning itself faster. It cannot accelerate the time it takes us to accumulate wisdom, make social meaning with each other, or deepen sensory understanding of our world. As long as we are human, learning will always happen at the speed of humans—consisting of feeling, thought, and movement—not at the speed of silicon.

The wheel extended our feet, the telephone extended our voice, and paper and pen extended the capacity of our memory (Clark & Chalmers, 1998; McLuhan, 1964). Now, hard drives and computers do the same. Even the spaces we live in and the groups we work with extend our thinking (Paul, 2021). Scholars have called these technologies "extensions of man" and the "extended mind." But *what is being extended* remains humankind and the human mind.

Striking a Balance

In a time of increasingly rapid technological change, we need something to ground our work in education. We want timeless, earned truths about learning that we can hold on to, and we want to connect those truths to emerging understanding about both technology and the brain. With this book, we aim to help build those connections. We affirm what the past has taught us, integrate what is new in the present, and offer our best effort at guidance for the future.

We also engage the fears and hopes we have heard from teachers and schools. Concerns about AI are many and real. Teachers and leaders speak with us about AI shortcutting learning and commercial technology driving school agendas, perpetuating biases, and posing ethical and environmental risks—and these must be addressed in our schools, across the education sector more broadly, and by technology creators themselves. Here, we discuss some of these concerns in terms of their historical context, the present trends, and the relevant underlying principles that can help guide our navigation of these difficult topics.

As we engage with these and other topics, we believe it is important to evaluate our prospective future with AI, not against perfect ideals—though those may be our guiding stars—but instead against our present reality. The question is not whether teaching and learning with AI is good or bad; it is whether AI improves the experience and outcomes of teaching and learning. In the context of bias, for example, the alternative to using a biased AI is not living in an unbiased world; it is living with biased humans (Greenwald & Banaji, 1995). Similarly, the alternative to AI providing ongoing but imperfect feedback for students is not teachers providing ongoing, perfect feedback for students; it is teachers providing a limited amount of imperfect feedback and often feeling burnt out from the hours they put into it. Today's concerns about AI are legitimate, but they must also be weighed against the reality of our current situation.

We offer this book as a blend of research, theory, and lessons from our experience about how to build the best human-centered, technologically enhanced world we can imagine. As you read, we want you to wonder about and engage in—as we do—what the future might look like for schools. We want this book to build our collective capacity to live up to that future. We want to think conceptually and creatively together so we are ready to participate in our ever-changing landscape.

And mostly, we want all to succeed—as teachers, school leaders, and supporters of teachers and school leaders—in service of our students.

Who We Are

We write this book from perspectives that span education, technology, leadership, neuroscience, and the arts.

After completing her master's degree in mind, brain, and education from Harvard Graduate School of Education, Maya spent ten years as a researcher and director at the Center for Curriculum Redesign, where she coauthored *Artificial Intelligence in Education: Promises and Implications for Teaching and Learning* and other publications. Since then, Maya has entered the classroom and taught middle school science for three years and is now working on her doctorate. Maya is also the founder of QuestionWell, a tool designed to maximize teacher agency and learning material quality with AI.

After fifteen years as a high school English teacher and school leader in the United States, Peter served as the head of school of King's Academy in Jordan. He designed an award-winning interdisciplinary digital humanities course, circulates a weekly newsletter called *The Educator's Notebook*, and is the cofounder of Athena Lab, a technology platform for teachers to connect and share practices. He serves on the advisory boards for the Center for Curriculum Redesign, SXSW EDU, and the Middle States Association's Responsible AI in Learning endorsement. He studied at Middlebury College and Columbia University.

In our various roles, we have been immersed in the science of learning and of AI, the development and implementation of educational technologies, the complexities of school leadership, and, most importantly, the daily study and practice of pedagogy—of teaching and learning in a classroom full of students. Through it all, we both also have engaged deeply in the arts: Maya as an improvisational actor and director and Peter as a musician and composer. Both of us agree that little captures the humanity and surprises of a classroom—the joy and the challenges—like improvisational art with other people. We have written this book through the various lenses we have worn as teachers, school leaders, researchers, scientists, technologists, and artists.

Who You Are

Who *you* are and your perspectives matter at least as much as ours—for how you read this book, what you may find valuable in it, and what you may disagree with. By "who you are," we don't mean your job or role—though we do provide specific action steps for different professional roles at the end of each chapter—but rather the values, beliefs, and philosophies you may have about education.

Our philosophies toward education influence the decisions we make when designing and developing schools, courses, units, and individual classes. Even how we respond to individual students in class is shaped by our educational philosophies: How do we encourage students to discover their interests? How do we focus on preparing them for jobs? How do we hold students to consistent and rigorous academic standards? Our answers to these questions depend on our beliefs about the purposes of education.

Since the late 20th century, researchers have sought to codify educational philosophies as *conceptions of curricula* or *meta-orientations toward education* (Cheung, 2000; Eisner & Vallance, 1974; Jenkins, 2009; McNeil, 1996; Schiro, 2013). We simply call them the *four teaching philosophies.*

1. **Academic:** Emphasizes rigorous intellectual training, disciplined scholarship, and mastery of academic domains
2. **Humanist:** Prioritizes student-centered growth, individual sensemaking, personal fulfillment, and self-actualization
3. **Social:** Seeks to address social goals through education and promotes civic engagement and societal transformation

4. **Pragmatic:** Focuses on efficient systems of teaching toward measurable outcomes, credentialing, workplace utility of knowledge and skills, and, often, technological solutions to school objectives

Notably, we are not limited to just one; each of us holds a combination of teaching philosophies (Cheung, 2000). Identifying our own teaching philosophies and those of others can help us pinpoint the framings, priorities, and underlying values we and our schools might be drawn toward or resistant to. Further, education-related conversations across different philosophies often surface different critiques of other philosophies (see table I.1). These may sound familiar from your own experience.

The significance of these differences comes into immediate focus when we talk about AI. Those with different teaching philosophies often focus on different opportunities and risks related to AI. Perhaps you have heard some common refrains in your conversations, too. Do colleagues get excited about the prospect of personalizing learning? Do they express concern about the environmental impact of AI data centers? Have they made the case that it is important to use AI to prepare students for the workplace? Or do they focus on fears related to academic integrity? Different responses often spring from different teaching philosophies. Table I.2 (page 10) organizes common responses to AI by teaching philosophy.

We share these teaching philosophies, critiques, and responses to AI as a means of mapping how schools or communities may be talking about AI. In the midst of each novel and uncertain moment in education's history, identifying our values—both individual and institutional—allows us to take a step back from the interpersonal friction that sometimes emerges in discussions about different visions for the future. As you move ahead, you might reflect on these questions.

- What are your school's prioritized educational philosophies?
- How might your school's philosophies shape its policies on adopting AI technologies in the classroom?

Table I.1: Teaching Philosophies and Critiques

	Academic . . .	Humanist . . .	Social . . .	Pragmatic . . .
Academic	**(self-perception)** . . . emphasizes rigorous intellectual training, disciplined scholarship, and mastery of academic domains	. . . is **too subjective**, lacks rigor and intellectual discipline, and sets outcomes that are unclear or vague	. . . is **too politicized**, risks ideological bias, and neglects foundational knowledge and skills	. . . is **overly mechanistic**, reduces learning to transactions, and neglects intellectual depth
Humanist	. . . is **detached from personal relevance** and neglectful of student-led sensemaking, emotions, and intrinsic motivations	**(self-perception)** . . . prioritizes student-centered growth, individual sensemaking, personal fulfillment, and self-actualization	. . . is **overly focused on external agendas** and prioritizes political or social ideals over personal interests and growth	. . . **is dehumanizing**, too rigid and impersonal, and ignores emotional, social, and creative aspects of learning
Social	. . . is **elitist** and overly traditional and ignores relevance to students' lives and pressing societal issues	. . . is **too individualistic** and neglects broader societal responsibility and collective good	**(self-perception)** . . . seeks to address social goals through education and promotes civic engagement and societal transformation	. . . is **too procedural** and disregards the critical and social consciousness necessary for addressing real-world issues
Pragmatic	. . . is **unfocused** and vague about measurable outcomes and overly abstract and theoretical	. . . is **too open-ended** and lacks accountability, structure, and measurable student achievement	. . . is **too idealistic** and impractical and lacks clearly defined and measurable objectives	**(self-perception)** . . . focuses on efficient systems of teaching toward measurable outcomes, credentialing, and workplace utility of knowledge and skills

Table I.2: *Teaching Philosophies and AI Opportunities and Risks for Teaching and Learning*

Teaching Philosophy	AI Opportunities	AI Risks
Academic Emphasizes rigorous intellectual training, disciplined scholarship, and mastery of academic domains	• Enables easier access to knowledge • Supports rigor through interactive learning experiences, ongoing feedback, and interdisciplinary connections	• Facilitates superficial learning and academic dishonesty • Weakens instructors' role as intellectual leaders
Humanist Prioritizes student-centered growth, individual sensemaking, personal fulfillment, and self-actualization	• Personalizes learning experiences to students' needs, interests, and progress • Frees up educators' time for mentorship and guidance	• Reduces interpersonal interaction • Reduces emotional understanding of others • Exposes personal student data
Social Seeks to address social goals through education and promotes civic engagement and societal transformation	• Prompts empathy and critical discussions about ethics and society • Enables social mobility for disadvantaged communities • Enables global awareness through immersive experiences	• Reinforces biases or misinformation • Reduces authentic engagement in communities and enforces top-down control • Causes disproportionate environmental harms to marginalized populations
Pragmatic Focuses on efficient systems of teaching toward measurable outcomes, credentialing, and workplace utility of knowledge and skills	• Improves job training and upskilling • Optimizes learning with personalization • Optimizes instruction with data-driven feedback • Improves efficiency by automating routine tasks	• Expands the gap between workforce needs and school curricula • Devalues credentials as a signal of student ability

- What is your personal teaching philosophy profile? How might that affect how you read this book?
- How might your responses to different chapters reflect a particular teaching philosophy?

We invite you to see both the opportunities and risks of AI in schools through the lenses of the various teaching philosophies discussed in this section.

How to View AI

This book encourages a foundational mindset shift for teachers and students: to treat AI less like a tool and more like an assistant, collaborator, or agent that works on our behalf. It can work with us and for us, and we can interact with it less like software and more like another person. We suggest this not because it's more accurate—we will discuss the issues in a moment—but because doing so will lead us toward useful ways of thinking.

We relate to each role—assistant, collaborator, or agent—differently. With an assistant, we set clear goals for what they need to accomplish, review their work, and provide feedback and ask for revisions. With a collaborator, we ask questions, explore possibilities, and build knowledge together. With an agent, we rely on them for a professional need and expect them to complete certain services.

We also recognize that assistants, collaborators, and agents make mistakes and have their own biases. We trust them with some of our information but not all of it—at least until certain boundaries and safeguards are in place.

AI is like this, too. It is a partner in our work, can serve in this range of capacities, and has limitations that lead us to engage with it prudently.

We do recognize that anthropomorphizing technology comes with risks. Treating AI like a person implicitly grants it agency, which can lead to giving up our own agency as humans, which can further lead to giving up our feelings of accountability and ownership (Shneiderman, 2022). This slippery slope leads to discussions that ask how long it will be until teachers are replaced by AI. We take these risks seriously and engage them in this book.

At the same time, treating AI like a tool also introduces risks, specifically when we consider what *we become* when we get in the habit of treating things that appear humanlike as if they are not human and exist just to serve our needs (Vidrin, in press).

For the sake of how the metaphor helps us get the most out of using AI—and for the sake of our psyches—we refer to AI more as a person than a tool. And for the sake of simplicity, we use the term *assistant* to encompass the roles of assistant, collaborator, and agent.

How This Book Is Organized

With AI positioned as an assistant, this book is organized around the different assistant-like roles AI might serve for teachers, students, and the shared classroom experience. To organize these roles, we outline three contexts where education happens.

1. Teachers working alone or in groups outside the classroom, researching and planning their classes and reviewing student work
2. Students working alone or in groups outside the classroom, learning and producing work
3. Teachers and students engaging together in the classroom

Each of these settings stands to gain from the participation of an AI assistant. In service of this, chapters in this book are built around the following assistant-like roles AI can play.

- **Part 1: AI for Teachers**
    - *Chapter 1*—AI is a **research assistant** for learning about our objectives, our subject, and our students.
    - *Chapter 2*—AI is a **planning assistant** for designing and customizing student learning experiences.
    - *Chapter 3*—AI is a **feedback assistant** for supporting teachers in grading and providing feedback for students.

- **Part 2: AI for Students**
  - *Chapter 4*—AI is a **learning assistant** for understanding new material.
  - *Chapter 5*—AI is a **doing assistant** for producing new work.
- **Part 3: AI for the Classroom**
  - *Chapter 6*—AI is an **administrative assistant** for managing classroom logistics and gathering data.
  - *Chapter 7*—AI is a **teaching assistant** for providing pedagogical insight during and after classes.

Together, these three contexts (teachers working, students learning, and teachers and students working and learning together) make up the settings of structured learning in schools. This book offers new ways to think about each of these contexts and leverage AI to enhance them. Further, as these contexts for learning all feed organically into each other, AI tools provide opportunities for more seamless and comprehensive communication and the transfer of information among them (see figure I.1).

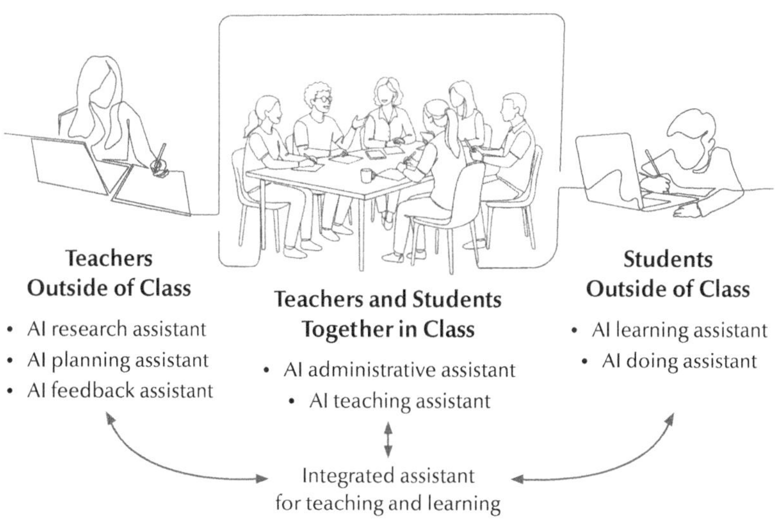

Figure I.1: *Ecosystem of teaching and learning with AI assistants across the three contexts of school.*

What We Mean When We Refer to AI

Before we move into the chapters on how AI can serve teachers, students, and classrooms, we first want to discuss what AI is and how we refer to it in this book.

The field of artificial intelligence was formally initiated at a summer conference at Dartmouth College in 1955. Since then, what we call *artificial intelligence* has evolved substantially as computer scientists have explored a variety of ways to simulate human intelligence. One of those approaches, inspired by how the brain works, was named *neural networks*, and it took particular root in the first decades of the 21st century. By the mid-2010s, AI technology could perform extraordinary tasks, from analyzing large quantities of data to generating new content from the data it analyzed. This technological breakthrough became known as *generative AI*, sometimes abbreviated to *genAI*.

Generative AI has opened entirely new opportunities for creativity and productivity by enabling machines to produce text, create images, and compose songs and videos that are generally perceived as humanlike. This breakthrough, especially because of the conversational human–computer interface it has enabled, has brought AI within reach of everyday people, not just computer scientists. Because generative AI is the means by which most people have come to consciously engage with AI tools, we have chosen to refer to generative AI tools simply as *AI*.

We also do not endorse any one particular AI tool. The landscape changes rapidly, and the general tools available—ChatGPT, Claude, Copilot, Gemini—may change. Similarly, education-specific tools range widely and evolve rapidly.

Where to Go From Here

The book builds cumulatively, but parts and chapters are also designed to stand alone. If you are interested in student learning, you might go straight to Part 2. If you are curious about how the teacher experience is changing, you might start with Part 1. If the classroom in an AI world is what feels most exciting, you might jump to Part 3.

Alternatively, you could go back a few pages and scan the table of contents. It intentionally includes headings from each chapter to offer a preview. Reading from beginning to end provides the most comprehensive immersion in the subject, but you also have agency in how you'd like to explore this book.

Citations are included throughout, in case you'd like to explore a topic further. You may choose to skip over them as you read, which we encourage for a smoother reading experience. A full list of our references, which span scholarship, art, journalism, and more, can be found at the end of the book for those who want to dig deeper.

Stepping into the future is almost always an act of improvisation. But not everything is up for grabs. AI has arrived (Reich & Dukes, 2024), but teaching and learning are not going to be "revolutionized." The very form and structure of this book reflect the assumption that AI will not fundamentally change education. The core experience of education is irreplaceable: It *will* and *must be* grounded in the relationships between students, teachers, and curricula. Whether in a classroom or in a virtual space, these relationships will remain.

Nonetheless, AI will be integrated into the contours of learning. The experiences of students, teachers, and curricula will change. How much they change is yet to be determined. This book aims to outline where and how some of those changes will happen—and have already begun.

Part 1

AI FOR TEACHERS

Great schools don't need buildings. They can function without administrators. They can work without technology, and they can lead students to prosperous lives without most of the things we see in schools every day.

But great schools cannot exist without teachers. Teachers are the heart of schools, so we begin our book here in recognition of their central importance.

Great teaching requires not only scholarship, empathy, discipline, and creativity but also high-level planning, scientific measurement, and responsive improvisation. Teachers master subjects, empathize with students' states of mind, adhere to objectives and standards, and devise learning experiences to match. Teachers also connect ideas, plan across a yearlong arc, assess progress daily, and repeatedly adjust both the daily and the yearlong plans in response to student mastery.

All this takes place across the three settings outlined in this book: (1) the work teachers do on their own as teachers, (2) the work teachers provide students to do on their own, and (3) the time teachers spend together with students in class. Together, actions in each of these settings create feedback loops for students and teachers alike. Here, in part 1, we explore the work of teachers.

Part 1 has three chapters that follow the course of what teachers do: They research in preparation for designing lessons, plan the lessons, and respond to what students produce as evidence of their learning.

- **Chapter 1, "The Research Assistant":** The experience of school begins taking shape when a teacher sits down to start planning or preparing their classes. But before (and while) drawing up a lesson plan, a teacher does research. This research is the focus of chapter 1, which explores how teachers research their curricular goals, their subject matter, and their students. This chapter explores the ways in which the human side of research—the passions that inspire teachers to work in schools day in and day out—can be strengthened by AI.
- **Chapter 2, "The Planning Assistant":** In chapter 2, we transition to the act of planning. Ultimately, all the research that happens in chapter 1—insights on content, curricula, and students—is channeled into designing learning experiences for students. The daily pace of lesson planning, we contend, makes teachers among the most creative people in the world, and these efforts to design and deliver learning experiences are enhanced and facilitated by AI tools.
- **Chapter 3, "The Feedback Assistant":** To close part 1, we explore how teachers manage student work. Providing thoughtful, personalized feedback for scores of students, week in and week out, is an unsustainable expectation for teachers. Teachers often must make a trade-off: the more compelling and higher quality their assessments and feedback, the more likely they are to face burnout. But rather than reducing learning to easily gradable but less meaningful forms of assessment and sacrificing nearly every waking hour to reviewing hundreds of student works and providing written responses to each, teachers can leverage AI to help provide relevant, timely, clear, and actionable feedback for students. This chapter explores mechanisms and strategies available now and what we expect to see in the years ahead.

Without further ado, read on!

CHAPTER 1

The Research Assistant

Student and Teacher Passion Inspire the Curriculum

If you want to build a ship, don't drum up the men to gather wood, divide the work, and give orders. Instead, teach them to yearn for the vast and endless sea.

—ANTOINE DE SAINT-EXUPÉRY

Before teaching, before walking into a classroom on day one, and even before drawing up a lesson plan, teachers do research. It is the first step in teaching and learning, and it starts with asking questions: "What are my curricular requirements?" "What content am I responsible for?" and "How prepared are my students?" These questions help teachers engineer a course.

But sometimes, teachers ask different questions entirely: "What do I want my students to learn?" "What content do I love to teach?" or "What topics do I think my students will find engaging?" These questions explore something else: the interests and motivations of the people in the classroom.

Is one of these approaches better than the other? The first looks structurally—*scientifically*—at the constraints of lesson design. The second looks personally—*artistically*—at the human passions and interests of the people involved. Where should teacher research begin? Should teachers follow the science or pursue the art?

On the one hand, lesson planning is a science. Careful studies have shown that certain pedagogical approaches work best for certain students under certain conditions. Also, research has surfaced underlying cognitive principles that have clear implications for lesson design. From these scientific approaches and principles, committees of experts have developed standards, sequences, and benchmarks for teachers of every discipline to follow. To meet these requirements—and often because of variability among teacher designs—some schools require teachers to follow fully-scripted curricula. Similarly, some schools enforce particular lesson planning templates or processes, believing that consistency driven by scientific research will lead to quality.

On the other hand, lesson planning is an art. Like songwriters, who might start their process inspired by lyrics, melody, or chords, teachers may start their process inspired by content, by students, or by requirements. Teacher *self-efficacy*—the degree to which they believe they can accomplish their goals—leads to more passionate, committed, and flexible teachers, more effective learning, and better school culture (Skaalvik & Skaalvik, 2007). From this autonomy comes the individual artistic fingerprint of each teacher.

You might see where this is going. Teaching is both science and art, but following the science and following the art seem to lead in different directions.

How do we reconcile this tension? Should the science of learning dictate a consistent way teachers research and develop their courses in service of quality? Or should teacher artistry and autonomy in service of motivation and inspiration guide their actions when they sit down to start researching? Are these goals—quality and autonomy—fundamentally at odds? Is it possible to create consistently high-quality lesson materials while granting teachers the autonomy they need to feel inspired and to inspire their students?

Maybe it wasn't possible to reconcile these tensions before AI, but in this chapter, we argue that with AI, it is.

The Marriage of Science and Art

Teaching is not the first discipline to blend science and art. Architects use physics and engineering to design functional buildings that are also artistically delightful. Filmmakers blend color theory and cutting-edge CGI to plan shots that achieve particular emotional effects. Industrial designers apply engineering and material science to create objects that are both useful and aesthetically pleasing. Even doctors rely not only on their knowledge of biology and chemistry but also on an understanding of bedside manner and an intuitive sense of what each patient needs. How might teachers do the same?

Science

The science of teaching and learning is increasingly well defined. Decades of research have now produced categories of what works in education that should all be parts of teachers' work. These include:

- **Evidence-based instructional approaches:** We know that explicit instruction with guided practice is especially effective for novices, while more open inquiry can be productive as expertise develops (Kalyuga, Ayres, Chandler, & Sweller, 2003); we know spacing, retrieval, and elaboration, especially as low-stakes practice with information-rich feedback, benefit durable learning (Dunlosky, Rawson, Marsh, Nathan, & Willingham, 2013); and we know prior knowledge, misconceptions, and cognitive load shape what learners can do in a given lesson (Bransford, Brown, & Cocking, 2000; Sweller, van Merriënboer, & Paas, 2019). These evidence bases offer foundations for informed approaches to instruction.
- **Discipline-specific knowledge:** Good teaching decisions are not generic—they hinge on the structure of the discipline and the representations that best reveal that structure to learners (Mishra & Koehler, 2006; Shulman, 1986). For example, research has shown that phonetic approaches are more effective than whole-word sight-reading approaches for teaching reading, and

visual representations of quantities are helpful for teaching early numeracy (Ehri, Nunes, Stahl, & Willows, 2001).

- **Standards and sequences:** High-quality standards and learning progressions offer scope and sequence that protect coherence across the curricula and equity of access. With this structure, each new learning objective builds on those that came before it. It does not accidentally skip a step, which would lose a subset of students who could not supplementally receive this knowledge from other settings, like at home or through tutoring (Schmidt & Houang, 2012).

Learning science has developed domains and principles of practice like these by studying the biological and psychological infrastructure of learning and its mechanisms. But applying learning science is easier said than done. "What works" is contextual. Claims made from research are based either on averages across diverse settings or on narrowly defined cases and are therefore difficult to generalize to new contexts. Translating science into practice is not a perfect science.

Art

Anyone who says that successful learning happens because of a coherent, standards-aligned curriculum is . . . correct. But that's also like saying that a successful visit to the library happens because of a reliable and well-designed car. Both are technically true but miss a key point. A successful visit to the library happens when a person with a functioning car *wants to get a book*. The car is the medium; the motivation makes it happen.

School is similar: The best learning happens when a teacher wants to teach and students want to learn. Passion matters. Research shows that the more passionate people are about their work, the more deeply they will invest in it (Ho, Wong, & Lee, 2011; Salessi, Omar, & Vaamonde, 2017). And in schools, the more committed and passionate a teacher is, the more they will inspire their students to have a similar passion to learn (Gilal, Zhang, Gilal, & Gilal, 2018).

A coherent, standards-aligned curriculum is important, but it doesn't help students much if the students don't care.

Science explores the way the world works, but art expresses and inspires the human spirit. Leaving individual inspiration out of the equation of teaching and learning overlooks the very things that make us human, like curiosity and motivation. Teachers want to bring more of who they are to their work. When 1,300 teachers were asked what activity they would most like to focus on if they had more time, more said individual planning and preparing for class (29 percent) than anything else—even more than actual teaching time (28 percent; Merrimack College, 2022). Many people tend to think about teachers bringing their whole selves to the visible part of teaching: when teachers are in front of students. But it turns out that even more so, teachers crave greater opportunity for ownership in the invisible parts of teaching: the design and preparation of lessons and instructional materials.

Research on intrinsic motivation affirms that autonomy is essential for fostering passionate, committed teachers who, in turn, foster passionate, committed students. Autonomy creates space for art—for teachers to draw on their intuition, taste, and individual understanding of situational nuance, not only for how to deliver class but also for how to design it.

While the science of teaching and learning ensures that a car has the capacity to take its passengers to the library, the art of teaching and learning is what fosters the motivation and passion to get on the road. Aligning them means reconciling forces that seem to pull in two different directions.

The Reconciliation of the Two

Combining science and art means navigating *constraints* imposed by science and *affordances* created by art. When architects build a bridge, they must abide by the constraints of physics, the particular location and materials, but that is not enough to design a bridge. To do that, they must have an artistic vision. The same is true for teaching. The science of learning establishes rules and principles that underlie successful learning. These are the *constraints* that provide requirements for learning design. They are limited in number and serve as a basis for planning. Meanwhile, the art of teaching invites new possibilities inspired by the uniqueness of teachers, students, and classroom content. These possibilities are *affordances* that open up new opportunities in learning design. They are as limitless as people are different.

It is important to note here that constraints, while limiting, are not inherently negative. Constraints provide necessary boundaries. Instead of being hindrances, they offer clarity (Brown, 2009). Artists, entrepreneurs, and others across creative fields have long made this point, from Igor Stravinsky (1947), who observed, "The more constraints one imposes, the more one frees one's self" (p. 65), to Marissa Mayer (2006), who states simply, "Creativity loves constraint." In other words, restrictions on creative work helpfully reduce the number of choices that need to be made, and for teachers, who make up to 1,500 decisions a day (Hattie, 2012), this can be a great benefit.

So, how do we both satisfy the constraints of evidence-based learning design *and* reach for the affordances of individual artistry? This is where an AI research assistant helps, and it happens when we start our lesson planning.

The Three Approaches of Lesson Planning

Do songwriters start with lyrics, melody, or chords? There is no one answer. Different songwriters start in different places at different times. Very few start with the science of sound, even though all are ultimately bound by its principles. Instead, most start with a feeling, whether it emerges as a tune, some words, or a chord progression.

So it is with teachers and lessons.

Educators are often coached to start their work by studying the *requirements* they must meet and what assessments they will use to measure success. Then, they work backward toward what *content* or materials they should use and what needs their *students* have. This linear approach is an eminently practical set of steps built on important curriculum-design principles (Wiggins & McTighe, 2005), and is depicted in figure 1.1. Like coaching a musician to study music theory before writing a song, it provides a solid foundation for teachers' knowledge of learning design. Without this guidance, curriculum design could get out of control quickly—or go nowhere—with the scope of a lesson or unit broadening or narrowing idiosyncratically.

But curriculum design principles, it turns out, aren't always the way excellent teachers start their research. In 2017, Peter led a team that invited state Teachers of the Year from all over the United States to illustrate their planning processes. It turns out that when planning their classes, even award-winning teachers start in a variety of different places. (See https://planningprocessillustrated.com for more about this project.) Some teachers start with standards, some with their students, and others with their own interest in what they're teaching.

Research has shown that while novice teachers tend to follow the materials they are given, experienced teachers do not plan this way (Kang, 2017; Kerr, 1981; McCutcheon, 1980; Pratiwi, 2024; Warren, 2000; Westerman, 1991). Instead, "experienced teachers' planning can be best described as a simultaneous consideration of . . . elements, rather than a step-by-step or linear progression of decision-making" (John, 2006, p. 488). Like musicians writing a song, experienced teachers start with an instinct, and it leads them through the essential elements of their learning design (see figure 1.2, page 26).

Just as a songwriter might start with lyrics, melody, or chords, a teacher might start with content, goals, or students. Each approach ultimately constrains *and* inspires the others. In songwriting, choosing a chord affects how the notes in the melody will sound; in teaching, choosing a learning outcome affects which materials will be appropriate and how they should be presented to students.

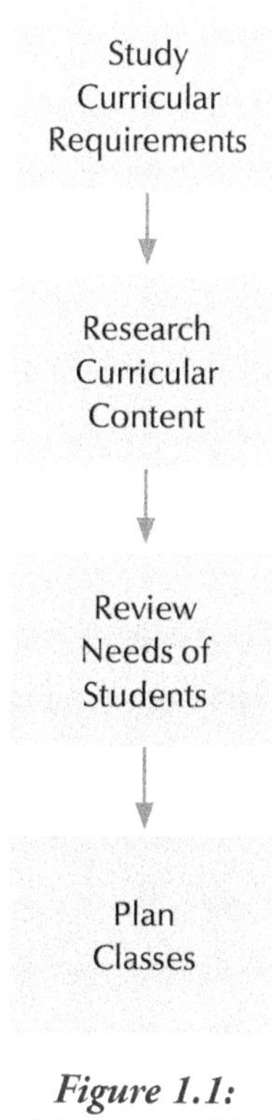

Figure 1.1: *A linear model of teacher research.*

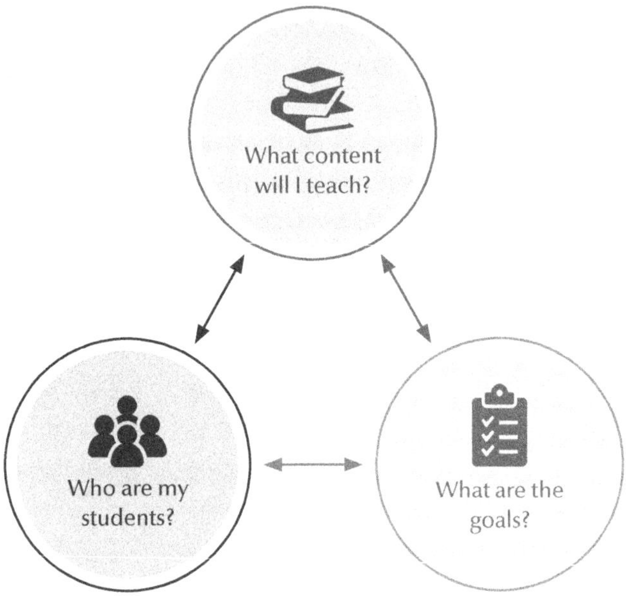

Figure 1.2: *The dynamic model of teacher research.*

When we consider content, goals, and students each in the context of constraints and affordances, we begin to see a new model for the many ways teachers go about planning a lesson. For each of the three areas—content, goals, and students—some aspects of lessons are predetermined, and others are the prerogative of the teacher. The predetermined elements are constraints, and they are represented in the center of figure 1.3. Teachers do not have a choice about these. The most powerful lessons, however, integrate affordances—those elements of a lesson inspired by the uniqueness of teachers and students. Affordances are what inspire teachers to move beyond the constraints, and they are represented by all the space outside the center of figure 1.3.

Crucially, our argument is that while it once may have been optimal to start with constraints, now, with AI research assistants, teachers have the capacity to start where their passion lies, make use of the affordances that come from doing so, and still be confident that they are meeting constraints. Teachers, like songwriters, may begin from where their inspiration lies, whether that is the content, goals, or students.

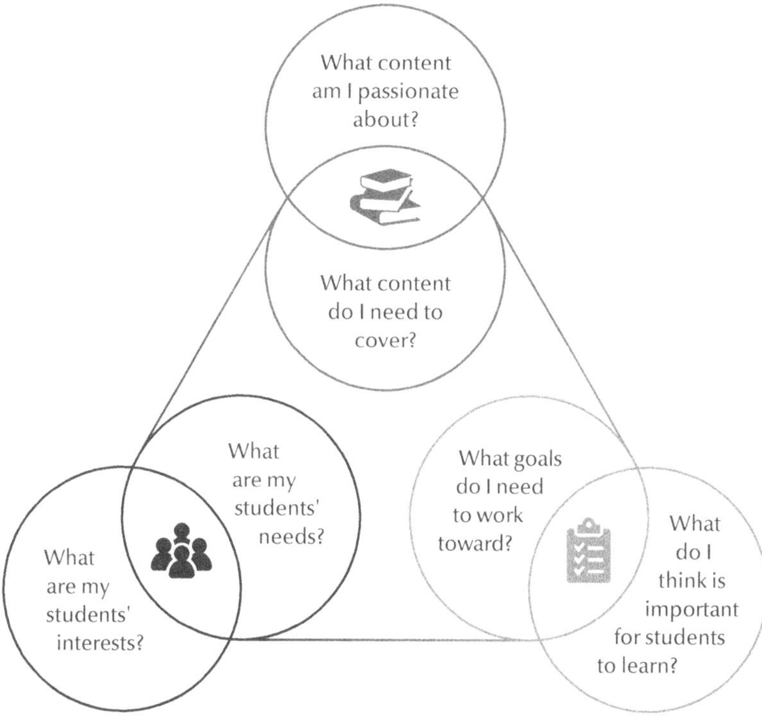

***Figure 1.3:** Constraints and affordances in the dynamic model of teacher research.*

Content First

Sometimes, teachers are inspired by the content and materials they teach with (see figure 1.4, page 28). When teachers light up from the clarity of a proof, the poignancy of a primary source, or the elegance of a well-designed lab, they show their passion for the content. When they know how to find just the right detail to frame just the right question, it often comes from a subject-matter passion that leads to knowing what resonates and what doesn't: that moment from *Romeo and Juliet* that works best for acting out in class, that particular insight about the women's suffrage movement that led to its success, or that one observation during a science experiment that turns the whole theory upside down.

In December 2023, Maya heard a story on NPR: the first CRISPR gene therapy patient was successfully treated for sickle cell anemia.

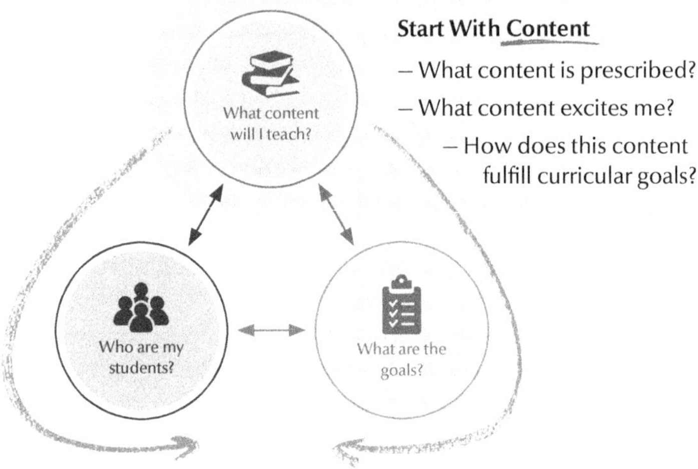

Figure 1.4: *Start with content.*

The interview with the patient was so moving and the content so relevant, she knew she had to play the clip for her class. She used AI to create guiding questions for the class discussion she led that day. Students were thrilled to listen to and discuss the course content in a way that was timely and relevant. This kind of passion for content leads to new discoveries and inspiration for teachers that translate to inspiration for students.

But how can we be sure that the planning that starts with content leads to learning experiences that also meet the constraints of well-designed lessons? This is where AI can be most helpful.

When teachers leverage AI as a research assistant to engage content, they can do so for two different purposes. One is exploratory: using it to expand their knowledge and collect fresh resources. The other is integrative: asking it to help bridge personal passions with the structure of the curriculum and students' interests.

In the first, teachers can use AI to deepen their content knowledge, whether to fill gaps or to keep up with what's new. To deepen knowledge, teachers can learn directly from AI, though AI's unreliability requires them to verify the information it provides. Another approach is to ask it for resources rather than trust it directly. AI can provide references to books and articles, links to videos, suggestions about podcasts, and more. Teachers can also use AI to find new content materials for their students. A teacher who has been given outdated content for a course could use AI to find more recent materials for their students, or they could use it to synthesize existing sources into an AI-generated text summary, podcast, or video, fueling the knowledge and interest of both the teacher and the students.

The second strength of AI is integrative. Teachers can use AI to connect the content they are passionate about with the constraints of curricular requirements and student needs. For example, suppose a teacher wants to read a particular book with their students. The teacher can gather both grade-level curricular requirements and information about their students' interests (perhaps from a start-of-the-school-year survey) and ask AI to make connections among them: "What elements of the book might best support course objectives?" and "How might the themes of the book be framed through students' interests?" Teachers can pose questions like these, review the AI-generated responses, remove what doesn't resonate or make sense, and add to their lesson plan what inspires them and serves the constraints of their class.

AI in Practice: Try this

Consider how you might deepen your disciplinary knowledge while aligning with your standards. Describe to AI an area of particular interest. Ask it to recommend readings on that topic and to provide an overview of possible avenues for further research. Verify that the recommended readings actually exist, then explore with AI how these readings might fill out your domain knowledge. The following are some example scenarios.

- You're a statistics teacher, and it's election season. Try asking AI what elements of statistics might be most relevant to elections. Ask follow-up questions based on your interests. When you have found a topic you find compelling, ask AI how that topic fits in with the AP Statistics curriculum and when in the year you might teach it. If you teach a different level of mathematics, try the same with your level.

- You're an English teacher, and you love a particular book that you think might be appropriate for your students. Try asking AI about the author or the context in which the book was written. Explore different social or cultural takes on the book. When you've found something of particular interest, try asking AI what standards the book might be particularly well suited to support. Be sure to verify them.

Learning Goals First

Teachers might start with required standards (see figure 1.5) for a variety of reasons. They may believe deeply in their importance, such as when the standards facilitate engagement with a pressing current issue or with an important life skill. Or they might be addressing a challenging subject, and the scope and sequence of requirements provide a crucial framework to ground their thinking. Or maybe they work with colleagues to stay aligned on the same set of requirements and compare results. When information about content and students is vast, requirements can be a productive framework to stay grounded

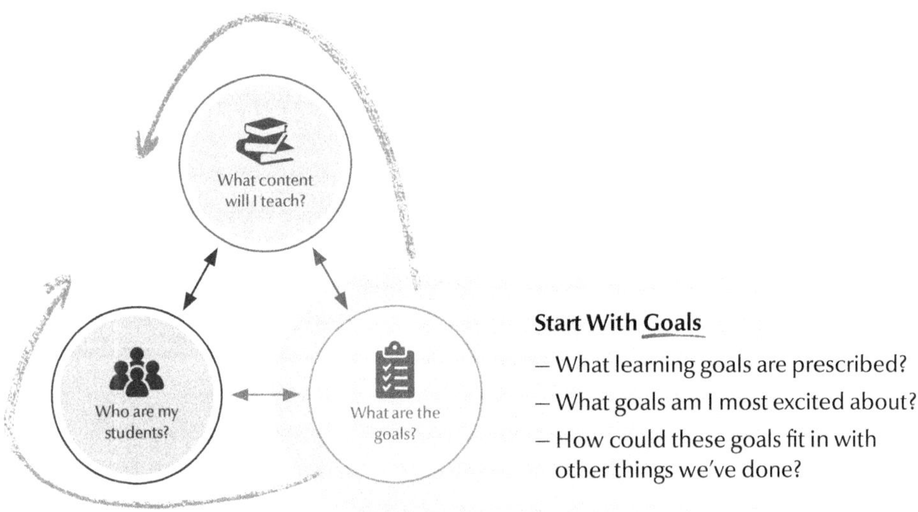

Figure 1.5: *Start with learning goals.*

and focused. Teachers may also appreciate the clarity that comes with a polished learning sequence of objectives that click together and build on each other.

What might it look like for a teacher to be excited about adding their own learning goals to required standards? When Maya noticed that her science curriculum included experiment design but didn't include confounding variables, she introduced the term and built it into lessons across the curriculum, prompting students to think more deeply about how to set up experiments and interpret results. Similarly, another teacher noticed students did not know how to use a notebook (start at the front and use each page sequentially), so she added this to her class's learning outcomes. Another teacher added email etiquette to her class's learning outcomes and directed her students to write a professional email to their parents every Wednesday that summarized what they did that week. In each case, the teacher's conviction that these skills matter led to new learning outcomes that complemented existing requirements and carried forward into students' broader growth.

AI in Practice: Try This

Asking AI questions about curricular goals and requirements might not sound like the most thrilling or illuminating experience, but trying the following inquiries might provoke new insights about what you're teaching.

- "What aspects of these goals tend to challenge students?"
- "I'm passionate about developing _____ in students. How might this naturally connect to the required goals for this unit?"
- "What kind of problem or challenge would naturally require students to use this skill?"
- "How might I rewrite this standard as an open-ended question?"
- "Which aspects of this goal could students personalize based on their interests, perspectives, or prior knowledge?"

As with a content-first approach, teachers can also leverage AI in exploratory and integrative ways if they begin with requirements. Exploratory uses invite teachers to dig into the standards themselves, while integrative uses help teachers link those standards to the content they teach and the students they serve.

Whether a teacher is excited about requirements or dreads them, they can use AI to explore requirements and deepen (or perhaps spark) their excitement by asking questions like the following.

- "Where do the skills learned in this lesson show up again later?"
- "What careers rely on them?"
- "What interdisciplinary connections might inspire students?"
- "What prior knowledge can teachers draw on given what they've covered with students so far?"

AI can review syllabi, lesson plans, assessments, and student work to extract learning goals that might be implicit in the work, and certain AI models can reason about how learning goals across a curriculum may be able to fit together.

AI can also help connect requirements to content and students in ways that broaden a teacher's choices. Which Next Generation Science Standards, for example, might be relevant for an upcoming lunar eclipse? Which mathematics standards could be useful for analyzing the statistics leading up to an election? How might an English language arts / literacy standard on argumentation connect to the op-eds students encounter in their social media feeds or to podcast debates they already enjoy? Which historical thinking skills could be sharpened by examining the narratives embedded in popular video games? Even smaller choices—like pairing grammatical standards with songs that have particularly complex syntax or aligning a unit on persuasive writing with a current school policy issue—can turn abstract requirements into tangible, resonant experiences.

By surfacing connections like these, AI can transform requirements from static lists into opportunities for creativity. Teachers who already find energy in requirements can use AI to deepen and diversify that excitement, and teachers who find requirements uninspiring can use AI to discover new angles that make the work more relevant and engaging for students.

Students First

When teachers are motivated primarily by their students, they are looking for opportunities to tap into student passions, interests, and needs. Teachers might begin with learning about their students or with content or activities that their students find inspiring. They might build connections between students' interests and course content and then shape learning experiences to fulfill curricular requirements, or the other way around (see figure 1.6).

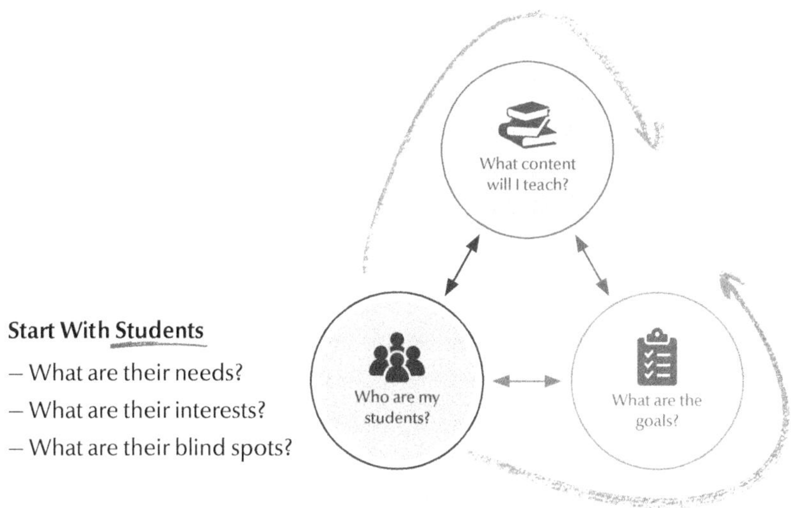

Figure 1.6: *Start with students.*

Perhaps a teacher would give a pretest to their students to see where they are starting and to motivate their learning by showing them gaps in their understanding. Or a teacher could have a class discussion where they collect information about students' interests so they can tailor the content to meet those interests.

When Maya was teaching middle school science, after seeing how excited Minecraft made her students, she created a project in which they would build a water treatment plant inside of Minecraft. She then figured out how to connect this student passion with the content and requirements for the unit and combined them into a rubric that students could use to make sure they were learning the material while enjoying Minecraft. The students asked about this project all school year and worked on creating the most elaborate structures they could (which still treated water with coagulation, flocculation, and so on)—all class, every class.

As when starting with content and goals, teachers who begin with their students can also use AI in exploratory and integrative ways. Exploratory uses help teachers learn more about who their students are, and integrative uses help teachers take what they know about students and connect it outward to content and requirements.

AI in Practice: Try This

- Survey your students about their favorite movies. Then, ask AI to suggest ways to connect the narratives of these movies to concepts in your subject (for example, the physics in a movie, the historical context of a movie's setting, or problem-solving strategies). Then, ask how these connections align with your learning objectives.
- If your students are passionate about a particular issue (such as climate change or health), ask AI to map that interest to standards or objectives in your discipline, then help you design a project-based unit that integrates the outcomes you seek to develop with the topic your students find interesting.
- Use AI to analyze themes across student work such as journals, reflections, or even quizzes, tests, and essays. Ensure that you have removed any personally identifying information. What recurring curiosities or frustrations emerge across the students' writing, and how might they be leveraged to shape lessons that feel both responsive and aligned with standards?

To help teachers better explore and know their students, AI tools can efficiently parse the reams of student data held in student information systems and learning management systems, quickly identifying areas of strength and areas for growth. What themes emerge from all previous comments from a student's language teachers? What subjects have historically been strengths and weaknesses for a particular student? What extracurricular activities have been most prominent in a student's past? What skills come up across subjects as needing particular improvement? What approaches have worked best in the past?

Especially in integrated information systems, AI-generated student profiles can highlight specific successes in the past or comments from previous teachers that indicate where current teachers might focus their energy or how they might most productively engage with each student. AI interactivity like this enables more effective—and easier—data-driven decision making. Note that using student information in AI systems should only be done in partnership with the school's IT office to protect the privacy of your students' data.

AI can also help connect student-driven ideas with content and requirements. If students are fascinated by environmental issues, AI might suggest connections to geography standards on human–environment interaction, mathematics standards on data analysis through climate datasets, or civics requirements on policy and decision making. Students interested in the culinary arts could explore science standards on the chemical reactions of cooking, mathematics standards on ratios and proportions, or health outcomes tied to nutrition. A class excited by robotics or maker projects might connect to physics standards on energy transfer, technology standards on design and iteration, or English language arts / literacy standards through technical writing. Even local community events—like a town festival or cultural celebration—could be linked to social studies standards, visual arts, or world language practice. In each case, AI helps teachers align what excites students with broader curricular goals.

AI in Practice: Classroom Example

Mr. Tennyson teaches twelfth-grade English language arts, and he loves Shakespeare's *Hamlet*. He has a role-playing idea in mind that he thinks will bring the text to life for his students. In it, students take the roles of characters from the play and stand in a circle around one student, who takes the role of Hamlet around act 3, scene 4. One at a time, they ask him questions about his character from other characters' points of view. Then, in a climactic moment, they all ask him their questions at the same time. The aim is to help students practice textual understanding and character analysis while developing empathy by exploring the many pressures Hamlet faces.

Mr. Tennyson feels pretty good about his mastery of the play, but he nevertheless thinks it would be helpful to brush up. He starts by seeking out additional critical views of Hamlet at this moment in the play, asking his AI assistant for two or three different perspectives on Hamlet's turmoil by the time he reaches act 3, scene 4. Mr. Tennyson also decides to share several lines from the scene with his AI assistant and asks for several different interpretations of those lines. The AI shares several insights into Hamlet's conflicted emotions, including one that Mr. Tennyson has not thought about in a long time.

Next, Mr. Tennyson considers the standards he is working toward. He knows that the exercise is meaningful for his students because they will engage in the text deeply and personally. But he has to make sure he is meeting course objectives, so he turns to his AI assistant to interpret the assignment and match it with relevant eleventh- and twelfth-grade English language arts / literacy Common Core standards. AI identifies several standards, such as RL.11-12.1, RL.11-12.3, RL.11-12.6, SL.11-12.1, SL.11-12.3, and W.11-12.9 (National Governors Association Center for Best Practices & Council of Chief State School Officers, 2010). Mr. Tennyson reviews them, omits one, and records the others as part of his objectives. He reflects on and notes how they align both with what students have done so far and what he has in mind for the future.

Then, to select the student who will play Hamlet, Mr. Tennyson asks AI to review his students' records to identify a student with strong analytical thinking, active participation, and previous interest in theater or performance. Pulling from courses the students have taken before, comments from previous teachers, and Mr. Tennyson's own notes, AI highlights two students who have demonstrated strong analytical skills, active participation, and a previous interest in theater or the performing arts. Mr. Tennyson knows that one of them hasn't had a win in a while and identifies this as an opportunity for a proud moment. He incorporates the selection of the student into his class plan.

Armed with a deeper understanding of act 3, scene 4, aligned standards, and an identified student for the role of Hamlet, Mr. Tennyson has finished his research and begins preparing the activity.

Review and Next Steps

What winter or spring day shapes teachers' lives the most? The day they learn which classes they will teach the following school year. And what day in late summer do they most anticipate? The day they learn who their students will be. These events give life to the year and set teachers on a course of learning about new content, requirements, and students. These days spark the research teachers pursue year in and year out.

Teacher research during lesson planning blends both science and art. While course standards and pedagogical principles provide a scientific framework for learning, artful teaching draws on a teacher's passion, intuition, and ability to inspire students. Historically, demands for consistent, high-quality lessons have created a tension between science and art, but in this chapter, we argued that AI can help bridge this gap, enabling both quality and autonomy by supporting teachers in the different aspects of lesson planning.

To do so, we identified three main approaches teachers take when beginning lesson planning: (1) content first, (2) learning goals first, and (3) students first. Regardless of the starting point, AI can enhance the process. For content-driven workflows, AI can deepen teachers' disciplinary knowledge and help integrate content with learning goals and student needs and interests. For goals-driven workflows, AI can help teachers explore standards more deeply and connect new insights to lesson content and student needs and interests. For student-driven workflows, AI can efficiently analyze student data and connect student needs and interests with both content and curricular requirements.

Further, with AI to support teachers in the dimensions of lesson planning they are least excited about, we have a new opportunity. Leaders can grant teachers autonomy and agency more broadly without being concerned about sacrificing lesson quality. The long-standing tension between an administration pushing for standardization and teachers pushing for autonomy and professionalism may have found a path toward resolution (Ingersoll, 2009). If AI can support teachers to lead with their passions and can complement their passions by connecting them to the other necessary research components, we can make sure that all lesson plans are high quality while being true to the teacher's vision.

But will we use this newfound freedom? In a famous fictionalized story, an elephant is tied to a stake when he is very young and cannot break free because the rope is stronger than he is. After trying to break free and failing, the elephant stops trying. But once he grows up, the elephant is large and strong enough that he *can* break free, yet he no longer tries and therefore stays tied up. We tell this story to demonstrate the idea of learned helplessness (see Maier & Seligman, 1976) and warn against it. For school leaders, teachers, and technologists, AI makes some previously impossible things possible. Will we take advantage of it?

What follows, here and in each chapter, are some next steps to consider for educators, school leaders, and those working in the technology industry.

For Educators

The following are some considerations for educators related to AI for lesson planning research.

Deepen your expertise in content areas you already know well. Suppose, for example, that you have recently studied the UN's sustainable development goals and have deepened your passion for environmental studies. Try exploring the topic with AI, diving into what you already know and what you're curious about. Ask it for sources to learn more. Probe it for open questions that might interest students. Follow your expertise to open up new avenues for discovery with AI. Draw on your knowledge. Be specific. Asking general questions will yield general results, while asking expert questions will yield more expert answers. This is a reminder that students in an AI age still need to develop mastery, as without it, they won't be able to discover as much or drive AI to deliver more. As you research your content areas, draw on your expertise to push AI further.

Explore an area where you don't feel confident. Conversely, if you feel you have gaps in your mastery, then leverage AI to help you explore and fill those gaps. Suppose you are an elementary school teacher, and you have never considered yourself a mathematics person. First, remember to have a growth mindset! Then, consider asking AI for different ways to think about simplifying complex fractions. What steps might make simplifying fractions easier? What are some varied examples that would help concretize your work with fractions? Similarly, if you have been asked to supervise the school debate team but don't know anything about debate, ask AI for an overview of middle school and high school debate competitions. What are the events? What are the criteria for success? How might students prepare? AI can help you close gaps you feel you have, whether those are in course content, understanding curricular standards, school activities, or many other things.

Explore your curriculum. Imagine you're ensuring that the content you are using in your course is aligned with the curriculum. How might you proceed? Consider uploading the standards you are responsible for, your school's course of study, or any departmental objectives. Consider including your school's Portrait of a Graduate or mission statement. Whatever the source material, share it with AI and engage in conversation with AI about the substance of these documents. Which areas do you most enjoy teaching? Which areas frustrate you? Share these with AI, and pose questions about them. How might you interpret your departmental objectives through the lens of your school's mission statement? How might your school's mission statement or Portrait of a Graduate accentuate some standards and objectives and diminish the importance of other standards and objectives? What skills in your discipline are not adequately represented in your school's documentation? Direct AI to create a table that correlates the standards and objectives you are responsible for with different words or phrases in your school's mission statement. Explore!

Consider your students' interests. What do your students love? Maybe it's a certain book, musician, or brand. Engage AI in a conversation about these topics to learn more about them. Then, explore with AI the ways in which these topics might connect directly, thematically, or conceptually with the content and standards you are teaching in your class. Who was that performer at the Super Bowl, and what was the theme? What is the current political firestorm, and how are students reacting to it? What is the movie that elementary school students can't stop watching, and how does it speak to the mission and values of the school? Let student interests shape your inquiries as you research.

For School Leaders

The following are some considerations for school leaders related to AI for classroom research.

Provide AI literacy training. For teachers to be effective guides for students, they must have strong AI literacy of their own. Include AI in routine annual professional development sessions. The principles behind responsible use of AI are not complex, but they require regular engagement to keep fresh (see page 132 for more on AI literacy). Teachers should have a sense of the mechanics, ethics, risks, and opportunities related to AI so they can foster similar sensibilities in their students.

Allow teachers to choose texts and topics. Teacher autonomy drives teacher satisfaction, which then creates commitment to student engagement. If you're not convinced after reading this chapter that teacher autonomy improves outcomes, then read Daniel H. Pink's (2009) book *Drive* for a deep dive into the research on and impact of intrinsic motivation. Ensure that the curriculum leaves space for and affirms the importance of teacher-designed learning experiences.

Schedule time for teachers to collaborate. The world is still figuring out how best to use AI for productive value. It will take decades to do so, especially since the field of AI is still undergoing rapid change and growth. We will be growing with this technology for years to come. Developing a portfolio of effective uses of AI won't come from one person but from many people developing many innovative use cases (Rogers, Singhal, & Quinlan, 2019). The more teachers can work together to develop, share, and refine the collective practices they make, the better the collective will become. Ensuring that teachers have ongoing, collaborative, practice-focused professional development will lead to the highest quality learning experiences for students (Darling-Hammond, Wei, Andree, Richardson, & Orphanos, 2009; Desimone, 2011).

For Technologists

The following are some considerations for those working in the technology industry related to AI for classroom research.

Support teacher learning, not just teacher work. Teachers love learning. Design tools for their learning, too. While edtech companies most often focus on what teachers are creating for students and on how they provide feedback for students, starting with what teachers want to learn and need to learn to feel successful at their jobs will improve the whole system. Help teachers master content and standards. Provide features that invite teachers to discover and learn for themselves. This will inspire us to engage more with your technology and provide even better learning experiences for our students.

Integrate with student information systems—or become one. When teacher research can pull on student histories, including previous grades, comments, and other data points, then they can have easy access to more holistic pictures of students. As in the example with Mr. Tennyson (page 36), information about students' extracurricular interests and previous performance will help teachers know how best to begin designing learning experiences.

CHAPTER 2

The Planning Assistant

Teachers Are the Most Creative People In the World

Dream a little before you think.

—TONI MORRISON

When we think of creative professions, we often think of the arts: painters, musicians, dancers, and more. Outside of the arts, the professions designated as "creative" are often those like designers, writers, and advertisers.

But teachers outperform all these professions when it comes to creative productivity.

Consider that advertising executives assign *teams* of professionals who take weeks or months to develop thirty-second experiences designed to influence viewers' buying patterns. But teachers create experiences *hundreds of times as long*, often by themselves, every single day, with an even more complex version of the same goal: not only to inform students but also to change their foundational knowledge and skills.

The magnitude of the work teachers do stood out to us immediately when we stepped into our respective classrooms. Preparing classes was similar to our experiences designing museum exhibits, developing theatrical performances, and creating immersive environments like escape rooms, except in teaching, we go one step further: The "audience" must be so affected by their learning experience that their approach to the world changes as a result. And unlike in other settings, students have not chosen to be there. This is a tall order.

Teachers' creativity spans categories. Teachers plan the arc of a yearlong or semester-long course like a novelist plans the three-hundred-page arc of a story. Teachers plan individual classes like a playwright shapes a scene in a play. Teachers design interactive learning experiences like a marketing agent designs a social media campaign. But also, teachers personally deliver these experiences like a musician or stage actor performing for an audience. They engage students like an improvisational performer taking suggestions from the crowd, inviting students to participate, and navigating the hecklers. And sometimes teachers remain silent as simple accompanists while the crowd takes over and sings for themselves.

Designing learning experiences that build habits and understanding in students is a *fundamentally* creative act. It is design work, execution, *and* improvisation. Teaching doesn't just tick the boxes of what we think of as creativity; it gets to the heart of what creativity *is*.

Lesson Planning as Creative Design

If teaching is a fundamentally creative act, then schools are not only *laboratories* for the science of student learning but also *studios* that support and foster teachers' creativity. And if lesson planning is a form of design, then teachers are designers—of experiences, relationships, and understanding. Understood this way, AI is a technology for advancing creative work. It amplifies brainstorming, prototypes lessons, and differentiates materials. It allows teachers to focus more on the right designs to support students and less on the mundane mechanics of their work.

This idea—that teachers are creative professionals—has a history in educational research, but it has only surfaced more recently in how the profession sees itself. The term *instructional design* started growing in educational discourse in the 1960s (Google Books Ngram Viewer, n.d.). And in the early 21st century, scholars explicitly connected the fields of education and design, referring to teaching as "design science" (Laurillard, 2012) and describing teachers as having "design knowledge" (McKenney, Kali, Markauskaite, & Voogt, 2015). Teachers develop, test, and revise models of learning, guided by feedback and reflection, like designers.

In other words, good teaching is more than content delivery; it is responsive, creative, and human-centered learning experience design. Planning is not just review and preparation; it is creative synthesis and construction. Creativity is not just a bonus feature of teaching; it is central to what teachers do when they shape curricula and respond to student needs. It is what enables teachers to more deeply imagine—and reimagine—what learning looks like every day.

We present teaching as the most creative profession in part because of our awe for and pride in what teachers do as a profession, but also because research and theory about creativity offer meaningful insight into how educators plan learning experiences. The following section looks first at the creative process generally, and then how each stage of teachers' creative processes more specifically can benefit from the support of AI.

The Creative Process

There are many ways to plan a class. Chapter 1 highlighted how teachers start their research in different places based on their source of inspiration: Sometimes they start with curricular objectives, sometimes with students' knowledge and interests, and sometimes with content knowledge and materials they're passionate about. Then, after research comes the creative process of planning: turning that research into learning experiences.

To make the creative process plainly accessible, we synthesized a variety of models of creative work into one description of the creative process that makes the most sense for teachers. This theory is primarily adapted from Charles M. Johnston's (2021) Creative Systems Theory and integrates models and language from design thinking (Brown, 2009), Donald M. Murray's (2003) writing process, the Osborn-Parnes Creative Problem Solving process (Parnes, 1966), and others (see figure 2.1, page 46). Distilled down, the creative process comprises five steps. Despite our unique planning processes, we all enact some version of the following steps.

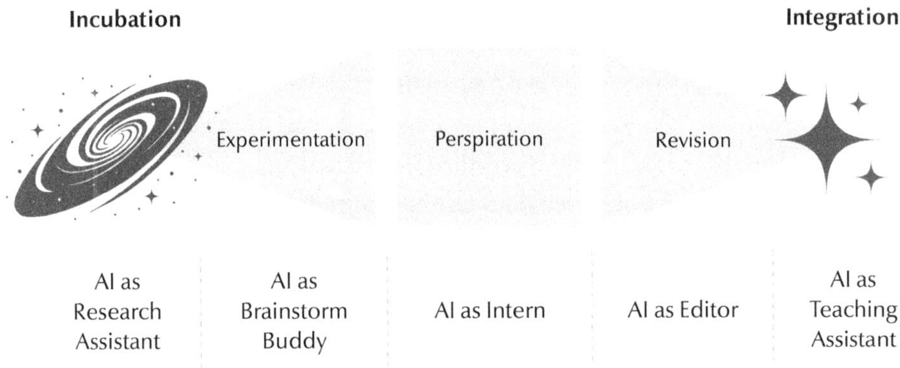

***Figure 2.1:** The Creative Process, adapted from Charles Johnston's* Creative Systems Theory.

- ***Incubation* of ideas while researching and exploring:** Before designing lesson plans, teachers research to learn about the content they will be teaching, their students, and the learning goals their courses aim to meet (see chapter 1).
- ***Experimentation* with ideas and how something new might take shape:** Teachers brainstorm or imagine different ways of shaping their class time based on their research. They explore different question prompts, pedagogies, activities, transitions, and more while taking into account the constraints of their classroom and the unique characteristics of their students.
- ***Perspiration* while doing the hard work of producing something new:** Teachers assemble an actual draft of their lesson plan, beginning to end, including any materials needed for the lesson. It isn't easy, and it takes time.
- ***Revision* of the final product of one's hard work:** Teachers make changes to their plans, sometimes on the fly, especially if they are fortunate enough to receive feedback from others.
- ***Integration* of what has been created into work and life:** Teachers implement the lesson given the day's constraints and opportunities and make connections to previous and subsequent lessons (see Part 3).

AI can support teachers at every step of this journey. The following sections explore how.

Incubation: AI as Research Assistant

Before we create, we learn. The more we learn and the more raw materials we have to draw from, the more we can create. If you haven't yet read chapter 1, we encourage you to do so for suggestions on the richly rewarding ways AI can support teacher passions during the incubation stage of the creative process.

Experimentation: AI as Brainstorm Buddy

During and after research, teachers begin their lesson planning by considering different ways to shape class. This stage is playful and experimental. Suppose, for example, you have found an article you'd like to teach. Will the students read the article alone and answer questions on a worksheet? Will they read it in small groups? According to what routine? Will they read it as a class and play a Kahoot! game afterward? Will they write a paragraph arguing for or against what the article suggests? Will they debate it? At this beginning stage of planning, we brainstorm and explore ideas, test them out to see which is right, imagine different possible outcomes, and expand and narrow our options.

AI is enormously helpful at this stage. AI tools surpass (most) humans in their ability to generate ideas (Guzik, Byrge, & Gilde, 2023). But humans are the final arbiters of the relevance of these ideas to their human goals (Faggioli et al., 2024). Together, we make a formidable team (Kasparov, 2010).

AI can help generate suggestions for how to teach a topic or text, including what media might be included, like articles, videos, podcast episodes, or lectures. Then, it is ultimately up to the teacher to craft the learning journey of the students.

AI in Practice: Try This

- Use AI to develop activity ideas for a topic you are passionate about. For example, Maya was teaching genetics but did not have a set curriculum, and she knew she wanted to spend some time on gene editing. She used AI to brainstorm ways to introduce the topic of gene editing and bring it to life. She ended up showing a short video guiding students through a hands-on, cut-out gene-editing simulation, then organized a class debate.
- If AI is being too generic, tell it to be more specific. For example, tell it to refer more directly to the text, try a particular pedagogy, or use a specific resource. Correct it when it is wrong. When Peter asked AI to suggest three ways to teach the introduction to *Paradise Lost*, its ideas were standard templates: read the lines aloud, offer historical context, introduce themes, and so on. He pushed back, asking for ideas that were grounded in the text's opening lines specifically. The results were much more interesting: begin with a discussion of religion framed by one of the opening lines, start with another well-known story that begins *in medias res*, unpack the specific allusions in the opening lines, and more. Push AI for more information, more detail, or more ideas. Tell it to be specific to your text or topic. Note that telling AI to be more "creative" or "original" will likely not be very effective. Like students, AI does best when it is given concrete, actionable directions.
- Think of a topic you enjoy teaching and would like to have a culminating project for. Tell AI what you teach in the unit in any format (include learning outcomes, describe activities you do throughout the unit, or upload worksheets), and then ask it to generate a menu of final project ideas for that unit. Consider feeding it additional constraints, such as "center the assessment on a relevant current event," "ensure the assessment fosters intrinsic motivation," "make the assessment a collaborative project," or "build iterative checkpoints into the assessment." Explore ideas that seem like they might be compelling by drilling down further into instructional steps, actions plans, rubrics, and more, or ask AI for different ideas.

Perspiration: AI as Intern

After experimenting with ideas, it's time to draft the lesson.

Once teachers have determined the objectives, content, and modality of a learning experience, the perspiration work begins: writing out the lesson and creating student-facing materials. This is often the most laborious and detail-intensive part of the process.

It is also where AI is used most often (Diliberti et al., 2024). AI can write lesson components in a variety of formats and include student instructions at different reading levels. It can write rubrics, identify relevant standards, and write questions that you can then export to the platform of your choice. Maya was so inspired by AI's ability to meet this particular need that she founded QuestionWell, a company whose first product specifically generates questions, letting teachers modify and integrate them into learning management systems and other teacher technology. In all these kinds of cases, AI will do the lifting once teachers decide exactly what and how to lift.

This takes on a special importance when classrooms have diverse levels of student readiness. Some classrooms require more differentiation, and producing differentiated materials takes time. Even creating more low-stakes practice opportunities is a way to add differentiation so students who need more practice can get what they need. Leveraging AI makes this differentiation possible.

We want to point out here that many calls to action with AI focus specifically on this stage of the teacher's creative process. "Teachers can use AI to produce lesson plans from scratch," edtech tools implicitly say as they market their easy-to-use tools to districts. We think this is only partially right. Without the deeper context and understanding that come from engaging with the first two stages of the creative process, lessons and materials drawn up at this stage are often generic and incomplete. Doing the work is less effective if time hasn't been spent clarifying and refining what work should be done.

AI in Practice: Try This

You've done the creative thinking, and now it's time to write out the plans. Share your thinking with AI, and see how much it can effectively and accurately produce for you. It's much easier to start with something (even if it's something flawed) than to start by staring at a blank page. Consider the following approaches to jump-start your perspiration work.

- **Language:** Instruct AI to produce ten verb-conjugation quiz questions in a format of your choice.
- **Writing:** Instruct AI to produce sample body paragraphs of an essay to demonstrate for students how a paragraph can have a strong topic sentence, well-integrated evidence, and meaningful analysis of that evidence.
- **Research:** Identify a research project that you teach. Instruct AI to write twenty possible topics for the research project. How many of them would you provide as possible topics for your students?
- **Science:** Instruct AI to create instructions for a lab based on your specifications of what the lab should do.

Each of these will likely require some kind of revision and customization. This might involve providing AI with what students are reading, the specific higher-order thinking goals you have in mind for a project, or the specific learning needs of your students. When you receive responses from AI, you may need to offer further directions about the format you would like the output to take: the structure of the page, the style or tone of the language, the subject matter in examples, or other details related to the AI's output. The more specifics you provide, the better the results.

Revision: AI as Editor

Draft in hand, it's time for revision, and like an on-call editor, AI is always available to provide feedback for you, the teacher. At a high level, what might be missing from a class or unit plan you have devised? What might you remove? Does the plan provide opportunities for both struggling and stronger students? Are there grammatical or typographical errors? Might you lay out the materials for a class differently? AI can help you by being both a critical and complimentary eye, surfacing potential strengths and weaknesses and suggesting different options.

Working through the perspiration and revision phases with AI is best done iteratively. A request to an assistant might not yield exactly the right response on the first try. In creative tasks, the best outcomes come from multiple iterations of divergent thinking (incubation and experimentation) and convergent thinking (perspiration and revision; Guilford, 1956).

Human oversight is essential. AI can produce work and provide feedback, but we must decide what we value, which AI outputs are useful, and to what ends. Even if AI has all the ideas, executes all the ideas, and critiques and improves all the ideas, the accountability for what to present to students lies with the teacher, embodied in every moment of judgment and decision making along the way.

AI in Practice: Try This

- Take an assignment you are *not quite happy with* and upload it to AI. Ask it what you can improve, and then select the improvements you would like (revision). Then ask it to make those improvements (perspiration).
- Take an assignment you are *already happy with* and upload it to AI. Ask it what you might improve, and consider the suggested improvements (revision). Then ask it to make the improvements you find compelling (perspiration).
- Take either assignment, and ask AI for an analysis of its strengths and weaknesses. Ask it to propose solutions for the weaknesses. Execute those solutions.
- Tell AI about your teaching philosophy. Do its suggestions change?
- Ask AI to make the assignment more accessible but also more challenging. How does it respond to seemingly contradictory requests?

Integration: AI as Teaching Assistant

Teachers' creativity must ultimately live in the real world, which is full of real pressures, tensions, constraints, and last-minute changes. Learning standards, for example, change at the speed of policy, in school or in government. Curricula, too, are often constraints that teachers work within that may change every few years as schools or districts adopt new textbooks or programs. At the yearly scale, students change, too, and teachers expand their repertoires of techniques as they attend professional development sessions, further adding constraints and affordances to the design process. There are also weekly changes, such as assemblies shifting the schedule around, and daily changes, such as fire drills and student sickness. Teaching plans also change in response to patterns in student work as well as feedback from students, coaches, and peers. Integrating our creative visions into the constraints and affordances of the day, week, year, and decade is its own step of the creative process. Teachers could use AI to brainstorm alternatives when they need to adjust their plans to the shifts in priorities and opportunities.

AI in Practice: Try This

- Take a lesson or activity that you've already planned, and upload it to AI, giving it specific real-world constraints you commonly face (for example, a sudden assembly cutting into class time, students needing additional review of a prerequisite skill, or a schoolwide initiative or theme). Ask AI to suggest adjustments that would make the lesson effective within these constraints. Review its suggestions critically, and choose the most practical ones.
- After adjusting your plan to accommodate immediate constraints, ask AI to help you map the adjustments with longer-term goals or standards. How can these real-time changes support or even enhance your broader curricular objectives? Analyze AI's recommendations, and select those that not only solve immediate challenges but also strategically align with your longer-term vision.

The Teacher's Creative Process in Practice

So what does AI-supported planning look like in practice? Well, that depends on which practice or pedagogy a teacher uses. Planning a seminar-based unit will look different from planning a lab unit or a series of lectures. Each pedagogy has its own "active ingredients" (Abry, Hulleman, & Rimm-Kaufman, 2015) that make it effective that teachers should focus their planning on—and AI can prove useful for advancing both planning and effectiveness.

For example, educators disagree about the role of explicit instruction in education. Some see it as lecturing and thus inherently dry and distant. Others point to the important factors that make the practice effective: clarity and structure in presenting new concepts, a focus on specific learning goals, the incorporation of multisensory aids, and an embedded formative assessment with immediate feedback to guide instruction (Toppo, 2025). Once these active ingredients are identified, the disagreements usually subside, and teachers can focus their attention on the elements of effective lesson design. They can then leverage AI to develop multisensory aids, brainstorm and clarify the presentation of new concepts, or devise formative feedback opportunities during explicit instruction.

For each of the pedagogical techniques in table 2.1 (page 54), we consider its most robust version when suggesting its use and brainstorming ways AI could be used to enhance it.

Teachers mix teaching modes to make learning experiences the way visual artists mix paints to make colors and musicians mix notes to make harmonies. For each of the educational strategies in table 2.1, AI can play a pivotal role not only in brainstorming ideas for engagement and understanding but also in creating, improving, and updating student-facing materials, teacher reference guides, and entire lesson plans.

In fact, early research suggests that teachers find the greatest productivity gains when they use AI not only to produce materials but also to help them brainstorm, learn, and explore possibilities before creating anything (Keppler, Sinchaisri, & Snyder, 2024). This suggests that while AI is useful in the perspiration stage for writing up

***Table 2.1:** AI Prompts Aligned With Teaching Techniques*

Pedagogical Technique	Incubation	Experimentation	Perspiration	Revision
Explicit Instruction	"Summarize recent research-based best practices for explicitly teaching [skill or concept]."	"Suggest three engaging analogies or metaphors for introducing [concept]. Explain the aspects of the analogy that do or do not align."	"Write a structured, explicit-instruction script for [number] minutes for teaching [concept], including formative assessment questions every [interval]."	"Review my explicit-instruction lesson on [concept]; evaluate its clarity, pacing, and formative assessment quality; and suggest specific improvements."
Guided Practice	"Find evidence-based strategies or examples of guided-practice activities that are effective for teaching [skill or concept]."	"Provide three scaffolded guided-practice activities that will help my [grade level] students learn [concept]."	"Create detailed student-facing instructions for a guided-practice activity on [skill or concept], including examples and clear checkpoints to assess student understanding."	"Evaluate the scaffolding, clarity, and student accessibility of my guided-practice activity about [topic], and suggest specific ways that I could better support student learning."
Independent Practice	"Identify recent evidence-based approaches for independent practice tasks that reinforce [skill or concept]."	"Brainstorm five engaging independent-practice activities that students can use to master [skill or concept] independently."	"Draft a differentiated set of independent-practice materials (worksheets, activities, or tasks) targeting [skill or concept], including answer keys and rubrics customized for [grade level or learning objective]."	"Critique and suggest specific revisions for my independent-practice assignment on [skill or concept], especially focusing on clarity, level of difficulty, and alignment with my learning outcomes."

Hands-On Learning	"Research recent classroom-friendly experiments or demonstrations related to [skill or concept] that are suitable for hands-on learning at the [grade] level."	"Suggest three hands-on activities or experiments that would effectively introduce or deepen student understanding of [skill or concept]."	"Write precise step-by-step instructions for students on conducting a hands-on lab about [scientific principle or topic], clearly listing materials, safety guidelines, and expected outcomes."	"Review and critique my student-facing instructions for a hands-on activity about [concept], suggesting improvements in clarity, safety guidelines, and practical implementation."
Project-Based Learning	"Find authentic, real-world issues local to [your area] that could form the basis of an engaging project-based learning experience in [concept]."	"Propose four detailed and creative project ideas for students to deeply explore [topic or unit theme], considering [diverse student interests and skill levels.]"	"Create comprehensive student-facing instructions, timelines, milestones, roles, and an assessment rubric for a project investigating [concept]."	"Evaluate the feasibility, clarity, and depth of the project description I wrote on [concept], and suggest specific improvements to enhance student engagement and accessibility."
Subject-Specific Strategies	"Gather recent examples of high-quality, challenging texts suitable for close-reading lessons focused on [literary skill]."	"Brainstorm three inquiry-based lab experiments for teaching [concept or skill] effectively in my [grade level] science classroom."	"Develop a detailed close-reading lesson plan with targeted discussion questions, annotated guides, and formative assessments for analyzing [literary text] by [author] in my English language arts class."	"Review the scientific accuracy, instructional clarity, and logistic practicality of my lab plan on [scientific phenomenon], and suggest precise refinements to my lab instructions and student handouts."

learning materials, it is just as, if not more, useful throughout our entire creative process. In other words, AI is best as more than just a time-saver; teachers report the greatest benefit when they engage with it as a creative partner.

Effective AI Practices Throughout the Creative Process

From our work with teachers using AI, we've noticed that several practices prove consistently useful. We'll go over them here, as they apply to every step of the creative process.

Break It Down

It's tempting to think of AI as able to complete a single task, even if that task is really a series of smaller tasks. While AI can often complete multiple tasks from a single command, it's important to remember that we can also string together multiple more detailed requests for more nuanced outputs. For example, ask AI to analyze an assignment for strengths and weaknesses and propose solutions for the weaknesses. Then pause, evaluate, and iterate the proposed solutions before executing them. As you are working, identify the smaller steps in the larger work you are performing and ask yourself, "Could I get to a starting point using AI? Could I pause and widen my thinking at each step, considering different ideas before committing to one idea?"

Try the Same Prompt but With Different Amounts of Context

The more detail you give AI, the more helpful it will be. Generic AI-generated lesson plans can sometimes be closer to an *outline* of a lesson plan than a lesson plan itself. The more AI knows about you and your students, the less generic and more tailored to your specific learners its responses will be. Tell it as much as you legally and ethically can, based on your school's established policies and practices.

It is important to never upload personally identifiable information—such as your or your students' full names, dates of birth, and so on—as some AI tools use the data you upload for the training of future models. Many terms of service of AI products explicitly ask you not to upload personally identifiable information, so by doing so, you would be violating multiple legal agreements: not only your school's policies but the software's policies. You may still ask AI to consider the important features of your class as a whole, however. For example, if there is a large portion of students who are multilingual learners, you can specify these general details to offer AI more context.

Challenge the Results AI Produces

AI sometimes produces false information—often called *hallucinations*. You will need to rely on your existing knowledge to challenge suspicious AI responses, especially if they seem too good to be true. These persuasive hallucinations are not a symptom that the AI is broken; however, they are a natural byproduct of how AI works. This is in part because AI tries to give the user what they want. It is important to remember that this tendency—pleasing people—is the standard of quality AI has been set to, not truth, which would be much harder to nail down, let alone train an AI model on. If AI did not hallucinate, it would not be able to create anything new—it would be, in essence, just a search engine.

AI also produces biased results that must be challenged. It is tempting to believe that technology has a neutral perspective. However, AI is not neutral. No technology is neutral. Words are not neutral, numbers are not neutral, and science is not neutral. In his book *The View From Nowhere*, Thomas Nagel (1986) argues that although humans aspire to a purely objective, unbiased knowledge detached from any perspective, position, and social context, this is not possible. It turns out that this is not only true of humans but of anything that humans create.

To illustrate this point, consider that an AI told that it is a car expert will respond in a way that is more knowledgeable about cars than one not told that it is a car expert or one told it was a bird expert. This seems benign, but an AI told that it was a man *also* behaved in a way that demonstrated more knowledge about cars than one told it was a woman (Salewski, Alaniz, Rio-Torto, Schulz, & Akata, 2023). This shows that AI is sensitive to perspective in nuanced and sometimes unproductive and harmful ways that are not immediately obvious. We can, of course, use AI's perspective-taking ability to our advantage, but to do it responsibly, we must be aware of the associated risks related to how AI works.

Challenge the results AI produces—not only because it hallucinates and has bias, but also to further propel its outputs in directions that follow your lead. Do not let AI replace your creativity; use it to realize *your* vision and goals. Be mindful of how and when you rely on AI, making sure you are still in the driver's seat. Trust yourself and your hard-earned wisdom.

Review and Next Steps

Teachers' work is creative work, and now every teacher has the support of their own creative team, ready to assist at every stage of the teacher's creative process. Each stage poses different challenges, and AI's role changes with them, from generating ideas (experimentation) to drafting materials (perspiration), providing feedback (revision), and helping teachers respond to shifting conditions (integration).

In all these stages, teachers use a range of instructional approaches: explicit instruction, project-based learning, guided practice, hands-on activities, and more. Rather than endorsing a single method, this chapter showed how AI can support any of them by generating aligned prompts, materials, and revisions that stay true to each strategy's active ingredients. For example, AI can help a teacher design scaffolded steps for guided practice, or it can brainstorm authentic issues for a project-based unit. By understanding the principles behind each technique, teachers can use AI to enhance the integrity of their favorite pedagogies.

Used like this, AI strengthens teacher agency. Like the best collaborators, it helps teachers think more imaginatively, design with greater ease, and refine their work with fresh perspective. It is a creative partner for the teacher acting as a novelist shaping a yearlong course, a playwright crafting a single class, a performer bringing energy to the room, and an accompanist stepping back when students take the lead.

When AI supports us in these ways, it enables our artistry, helping us create inspired and inspiring learning experiences for students.

What follows are some next steps to consider for educators, school leaders, and those working in the technology industry.

For Educators

The following are some considerations for educators related to AI for classroom planning.

Try the prompts in the AI in Practice sections. As educators, we know there's nothing quite like experiential learning. Try the suggested prompts throughout this chapter for yourself. Try them in different ways. Set a goal, like trying ten different approaches to a problem you have, and see how far you get. Push AI to go further, even if it's with generic prompts like "How would you improve this?" and "Implement those changes." Maybe you'll have ideas for new avenues to explore. Share your findings with your colleagues to get the most out of the experience.

Identify more subtle moments when you could use AI in your planning. As you prepare for your next lesson, notice the moments when you take a deep breath to prepare yourself to dive into a focused work mode. Could AI help lighten the load of that work? Notice moments when you need more ideas. Could AI help you brainstorm? Notice moments when you feel there is a way to improve what you're doing, but you're not sure how. Could AI provide an analysis? Add a reminder to help you get used to reaching for AI in a time of need. Ideally, it becomes an ongoing process in the back of your mind, similar to how a teacher has a feeling that it's time to throw in an activity to wake students up, or a researcher has a feeling that they need to back up a certain statement with a citation. It becomes an ongoing process that chimes in when necessary.

Gather a practice group and compare notes. The magic of working in a group is that when you are stuck, someone else might help you get unstuck and vice versa. When engaging in such a creative exploration, it's important to learn from as many different vantage points as possible. If you notice differences in your approaches, interrogate them to see if there are differences in underlying beliefs, values, understanding, or simply in preferences.

For School Leaders

The following are some considerations for school leaders related to AI for classroom planning.

Identify pain points for your teachers. Gather information on your teachers' most pressing needs while being mindful not to take up more of their limited time. You can learn directly in conversation with teachers or through other means that don't take more of their time. This can be done during routine observations or conversations, faculty meetings, or professional development days. It's important that teachers work on real, tangible problems rather than try to solve potential problems that might only serve to add to their workloads.

Allocate time for teachers to explore AI collaboratively with colleagues. It can be very challenging—but very rewarding—for school leaders to give teachers time to work collaboratively with their colleagues (Patrick, 2022). Design this work time like you would a class. For example, you could do research and choose ahead of time a handful of AI apps that teachers might explore. Then, they could work in their departments or teams to create or rework a lesson together. Or you could set a specific goal for each department and let them explore different tools. How you design it will depend on what learning goals you have for your teachers. Design with them.

Offer professional development that gives teachers space to adjust their lessons. How often do teachers walk out of a professional development session feeling that their time was well spent? When there are pressing concerns, such as preparing for the next day, it can be hard to suspend that pressure and focus on an abstract idea. Use teachers' motivations instead of working against them. We know that students learn more deeply when they have time to play around with ideas with their peers on a topic they are passionate about, yet we sometimes deny this when it comes to designing learning experiences for teachers.

For Technologists

The following are some considerations for those working in the technology industry related to AI for classroom planning.

Uplift teacher collaboration. The more teachers can collaborate with each other while using your technology, the more they can collaborate, period. Whether it's sharing their creations or working together to create them, collaboration is immensely useful to teachers and attractive to schools. Research has shown over and over again that effective professional development is collaborative, practice focused, and ongoing (Darling-Hammond et al., 2009; Desimone, 2011). If a school is trying to promote collaboration in professional development, it will be grateful if your tool helps facilitate that collaboration.

Design for interoperability with other tools. While some start-up wisdom has dictated that it's best to get users hooked on your tool because everyone else is a competitor, this is not how educators work. Rather than competing with anyone and everyone, see if you can solve a specific problem for users and support them in using other tools to solve other problems.

Guide teachers in the creative process. We know that AI works best when it is given guidance, but many users will not automatically give AI the guidance it needs. Therefore, create a user experience that elicits the guidance that would result in the best outcomes. For example, if it would help to know more about who is in a class to give pedagogical suggestions, add a question to the workflow where teachers describe who is in the class. One or two added questions can make the difference between a high-quality, usable AI tool and one that is just a gimmick.

CHAPTER 3

The Feedback Assistant

AI Should Not Put the Final Grade on Anything, But It Should Provide High Quality Feedback On Everything

When students receive both scores and comments, the first thing they look at is their score, and the second thing they look at is . . . someone else's score.

—DYLAN WILIAM

To many students, grades are what give value to their work. They may see grades as the judge, jury, and end-all-be-all determiners of their futures. The problem with this is that when students focus on grades, they miss the feedback that is crucial for growth.

This is obviously not what teachers want. Teachers want students to engage with formative feedback. To teachers, assessments are learning opportunities—*essay*, after all, comes from the French *essayer* (to try). An essay is just one of many attempts at growth, and that's where the learning is: the cycle of trying, getting feedback, adjusting, and trying again, better informed. Some teachers, if they could, would *only* offer formative feedback to serve this cycle of growth. But grades and summative scores do have their purposes. They can show how students are doing at a glance, they enable teachers to surface trends across a class, and they can become the basis for selecting and developing interventions and next steps. Unfortunately, these same summative grades are often what students pay attention to the most.

Fixation on grades happens in part because grades have permanence on a transcript. As a result, grades can be decisive in students' futures, such as in how students are placed in courses the following school year, whether they qualify for admission to a college, and, in some countries, even whether some professions are open to them. In other cases, grades can lead to praise or consequences at home, qualify or disqualify students from scholarships, and determine whether students are able to participate in sports. These downstream outcomes from grades put pressure on students to focus more on the score than on the feedback.

So, teachers go through all sorts of contortions to refocus students' attention back on feedback. They put comments on the front page of an assessment and the grade on the last. They omit the grade from the assessment altogether and require students to write responses to written feedback before they can see their grades. They even make up different forms of grades, like one of Peter's teachers, who drew a bunch of grapes and a stick-figure hand different distances away from the grapes on the back page of each essay to show how much students' writing "grasped" a big idea. That teacher wanted his students to think about the idea, not the mark. What did Peter's classmates do instead? They took out rulers to measure, to the millimeter, the teacher's evaluation of the quality of their work. For valid reasons based on the systems they are in, many students see grades in this way—as quantified statements of judgment and worth, not as part of the growth experience itself. But as Peter's teacher sought to convey, it's the reaching that matters for learning, not how far away or how close a student might be at a given point in time.

Grading and assessment—distinct from feedback—are enormous topics worthy of books in themselves, and they are the subject of much consternation and innovation. Perhaps one day, students will not have grades, and they will be able to learn from ongoing, unmediated feedback—and perhaps that day will be sooner than we think—but in the meantime, we must try to provide as much appropriate feedback for learning as possible.

Still, providing feedback takes time, and consistently and sustainably offering lots of feedback is one of teachers' biggest challenges. The more detailed and personalized the feedback, the longer it takes to produce. Multiply the time it takes to provide feedback for one student by the number of students, and suddenly, dozens of tests or essays become hours and even days of teacher work. Maintaining a regular cadence of nuanced, personalized feedback for every student quickly becomes untenable.

However, with technology available to provide feedback immediately and at any hour, students have additional support throughout the learning process. Now, students don't need to wait for feedback on assignments or even their day-to-day work. AI can provide initial, first-pass feedback on students' attempts, students can incorporate that feedback, and AI can provide as many further rounds of feedback as the student wishes until the student is ready to submit their work to their teacher. Teachers, relieved of the need to review an insurmountable amount of work, can be selective about what feedback to delegate to an AI assistant and what feedback to provide themselves, opening up the floor for deeper conversation and relationship building.

But is feedback from AI assistants effective? Or, rather, *in what ways* is AI feedback effective? And in what ways is human feedback better? These are the focal points of this chapter.

Elements of Effective Feedback

Gobs of good, consistent feedback is essential.

—RICHARD HERSH

Humans and machines both learn through simple three-step experiences: (1) They take actions, (2) they receive feedback, and (3) they adjust their next actions. This cycle repeats: actions, feedback, adjustments, new actions, feedback, adjustments, and so on, as shown in figure 3.1 (page 66).

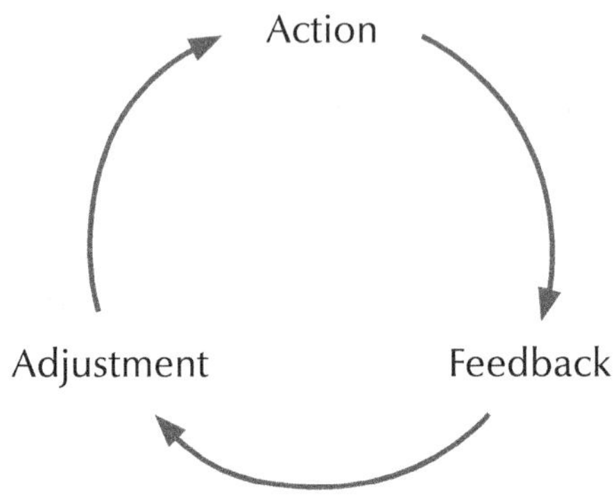

Figure 3.1: *The action-feedback-adjustment loop.*

This is a feedback loop, in which an output is "fed back" as an input, and the cycle repeats. It's called a *feedback loop* because of that crucial step: Feedback from the outcome of one action fundamentally informs the next iteration. This is how humans (and other animals) have been learning for millennia: act, see the consequences of those actions, adjust, and act again.

This is also how learning works in many skills-based professions: Feedback—and thus learning—is immediate. A chef tastes their soup throughout the cooking process and decides whether to add more salt (or add less next time). Their learning is clear, immediate, and ongoing. The taste provides them the information they need to evaluate their previous actions, adjust, and get ready for future actions.

But feedback in schools tends to be more challenging. Unlike chefs, students rely on other people for feedback that can help them see how they achieved or failed to achieve the learning objectives they were aiming for. The ultimate goal, however, is for students to be like the chef: to have the knowledge of what makes a good soup, when and how to taste their soup, and how to isolate the different important dimensions of its taste so they can take action and adjust. The point, again, is to set students up to independently exercise their agency so they can carry that with them outside the classroom.

Machines are not so different. For them, feedback is often built into their learning processes. Most machine learning begins with large datasets that are akin to flash cards. Humans might provide a machine with a set of images, for example, and each image would have a label. The machine then makes a guess about each label, compares it to the correct answer, and adjusts its parameters to improve future guesses (Rumelhart, Hinton, & Williams, 1986).

Like a human who does mathematics homework and then checks answers in the back of the book, the machine verifies its work and refines its performance over time. Unlike humans, though, machines can churn through millions of these feedback cycles in just hours or days. And machines aren't limited to images—they can learn what sounds match what words, what words are most likely to follow other words, and so on. This is, in fact, how AI learns how to speak, so to speak. It is remarkable that the power of generative AI that we see today is built on a system receiving simple feedback on whether it was correct or incorrect.

For students—who cannot churn through millions of feedback cycles—this kind of correct-or-incorrect reinforcement feedback is not as effective. Offering a correction leads to slightly more learning, but the most learning comes from high-information feedback (Lee, Peeters, Auer, Giesbers, & Appels, 2024; Wisniewski, Zierer, & Hattie, 2020).

For example, if a student makes a mistake, asserting that plants get their energy from the soil, a teacher providing reinforcement feedback would simply say, "Incorrect," while a teacher giving corrective feedback would say, "Plants get their energy from the sun, not the soil." But a teacher offering high-information feedback might say, "I see why you might think that plants get energy from the soil—because they do absorb nutrients from it. Think about the equation for photosynthesis. What is the source of energy in that process? Try explaining it in your own words." This kind of actionable, high-information feedback has shown a far greater effect on student performance across many studies (see figure 3.2, page 68).

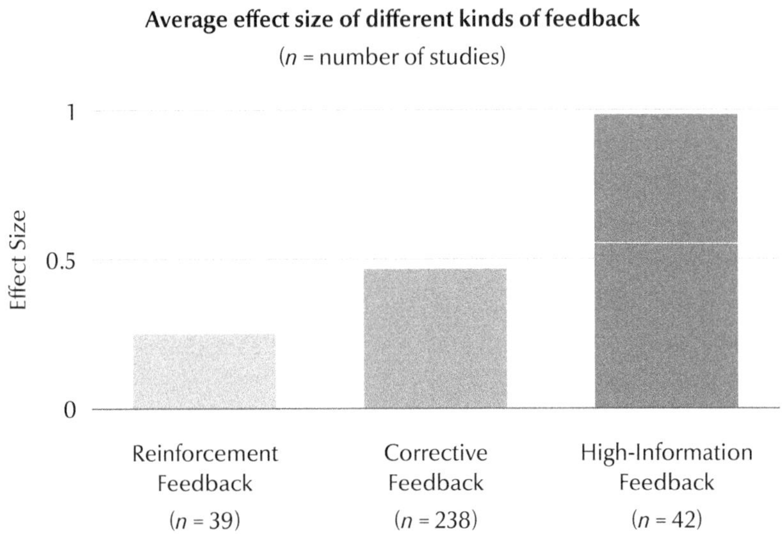

Source: Adapted from Wisniewski, Zierer, & Hattie, 2020.

Figure 3.2: *The effectiveness of high-information feedback.*

To figure out whether, to what degree, and how AI could help with student feedback, it is important to first define effective feedback clearly and simply. In a seminal essay, educator and author Grant Wiggins (2012) identified seven characteristics of effective feedback: Effective feedback is (1) goal referenced, (2) tangible and transparent, (3) actionable, (4) user friendly, (5) timely, (6) ongoing, and (7) consistent. The first descriptor (goal referenced) has to do with the *relevance* of the feedback and the final three (timely, ongoing, and consistent) with its *timing*. The second and fourth (tangible and transparent and user friendly) have to do with the clarity of the feedback, while the third (actionable) describes the overall utility of the feedback for student agency. For our purposes, Wiggins' characteristics of effective feedback might be distilled to the following three elements that support student agency.

1. Relevance
2. Timeliness
3. Clarity

In this chapter, we explore how AI advances our capacity to provide effective first-pass feedback that satisfies all three characteristics.

Relevance

Peter knew a new art teacher who once pulled an essay written by one of her students out of her shoulder bag. She had filled nearly every centimeter of white space on the paper with comments. Marks from her green pen blanketed the margins and the spaces between the double-spaced lines, the top and bottom of the pages—everywhere. Her good intention had been to provide as much feedback as possible, but the outcome was overwhelming, even for her. She had gone to her mentor for guidance on how to coach this student, who she felt was failing.

"What was the objective of the assignment?" her mentor asked.

After the teacher responded, the mentor followed up: "How much of the feedback relates to the objective, and how much relates to other issues?"

After talking with her mentor, it turned out that most of her feedback was about shortcomings in the writing that weren't directly related to the goals of the assignment. These may have been important for the student's growth as a writer, but they weren't immediately relevant to the work at hand. This realization was liberating.

Teachers don't need to correct every error on every assignment every time. When students submit draft lab reports, teachers might focus their feedback on the procedures or data. When students give a practice presentation, their feedback might focus on tone of voice and not spelling mistakes. When students freewrite or share journal entries, teachers might provide feedback on the ideas and not the punctuation. Finished products might require broader attention, but again, only to the degree that suits the present objectives. Feedback that isn't relevant to instructional goals can be distracting to learning—sometimes even inhibiting learning (Hattie & Timperley, 2007).

For a chef to eventually know what to do next just based on a teaspoon-sized sip of soup, they need to understand what they are tasting *for*. Anyone can taste soup and report whether they enjoyed it. Knowing what to do based on the taste involves understanding what the goals are and which actions will have which consequences on those goals. Tying feedback directly to the objectives of the assignment helps the student build up an understanding of their past actions and thus a model for their future actions.

But this poses a question: If teachers can sometimes lose track of what is most relevant, how can we expect machines, which have a limited understanding of the context of students' lives, to stick to what's most relevant?

Several simple practices could make gaining relevant feedback from machines feasible and scalable.

- **Provide AI with the objectives and criteria for a project or assignment:** Upload an assignment sheet or learning objectives to an AI tool. Consider referencing standards that the project or assessment aims to address. These documents and references provide lenses through which AI can view student work. Then, ask AI to assess how effectively a draft achieves the objectives or meets criteria.
- **Upload a rubric:** Rubrics create a consistent framework for assessing student work and communicating standards of performance. Using these frameworks, AI can provide contextually relevant and focused feedback, approximating the quality and specificity of human feedback (Escalante, Pack, & Barrett, 2023; Kurt & Kurt, 2024; Li et al., 2024; Polakova & Ivenz, 2024; Venter, Coetzee, & Schmulian, 2025; Wu et al., 2025).
- **Leverage AI to develop the rubrics themselves:** To help both the student and AI, teachers might specify their students' grade level, the course, and the nature and objectives (or standards) of the work (see chapter 1). They can then prompt AI to develop a rubric with multiple criteria, including clear descriptions and examples of mastery at different levels of performance. Developing effective rubrics will likely take several iterations, but once complete, these rubrics serve a dual purpose: providing students with models of what strong work looks like and giving AI a reference point for providing feedback on student submissions.

Together, these practices ensure that AI feedback is grounded in the same learning goals that guide teachers and students. Tying feedback directly to learning goals ensures AI stays on topic, minimizes the cognitive load for students interpreting the feedback, and clarifies how those objectives can inform their future actions.

Timeliness

A teacher's latest nights are their grading nights. Working through piles of tests or essays takes hours, and reviewing batches of student work to return them in a timely manner is among the most time-consuming tasks for teachers (Merrimack College, 2022). Sometimes, we succeed, but often, it isn't possible. Nearly every student everywhere asks at some point, "Will we get our tests today?" The worst creeping guilt we personally felt as teachers came on the days when we hadn't yet returned students' quizzes, tests, or essays in the time we intended.

Even when student work is consistently returned swiftly, the timing could still be better. The simple fact is that feedback almost always comes *after* the work is done, and learning happens best *while* the work is being done—while interpreting the data, writing the essay, solving the problem, reading the source, or cooking the soup. This is one of the reasons why one-to-one tutoring is so effective—the most effective feedback is timely, delivered in the moment, when students can adjust their actions, see the consequences, and close the feedback loop that leads to learning (VanLehn, 2011).

But giving in-the-moment feedback for every student all the time isn't possible—at least for humans. So most teachers end up giving most feedback on final *products*, when what they want to do—and what is best—is guide students throughout the *process* of their learning. As a result, the feedback teachers give can feel more like summative evaluation and less like guidance or coaching.

With AI, however, continuous, formative feedback for all students is possible. Indefatigable and always available, AI feedback assistants can support the full arc of student work.

Here, we explore what AI-facilitated feedback looks like through the first four stages of the creative process we described in chapter 2, but this time applied to *students'* creative work rather than teachers'. In particular, the first four stages of the creative process are ripe opportunities for feedback.

- **Incubation:** Feedback during the exploration stage of student work might introduce new sources for input to students' thinking.
- **Experimentation:** Feedback when generating early ideas assesses emerging possibilities in student work and guides students toward the most productive or generative paths. It can help students explore different approaches without having to commit to them.
- **Perspiration:** Feedback while doing the hard work—such as when practicing words in a new language, writing pages and paragraphs, or iterating a slide deck—supports granular-level improvements.
- **Revision:** Feedback after students have completed a draft—of a lab report, a solution to a mathematics problem, or an essay, for example—helps students look back at the work as a whole, assess their success against relevant criteria, and reflect on their process as well as their product.

Feedback during different stages of student work will look different in different subjects. For writing, during the incubation stage, when a student brainstorms topics or arguments, an AI assistant might suggest relevant sources or questions. During the experimentation stage, when students share initial ideas or outlines, AI might provide feedback on their viability, relevance, or effectiveness in relation to assignment objectives. As students move into the perspiration phase and draft paragraphs or sections of an essay, AI might affirm decisions that align with objectives, offer corrections where needed, or help them overcome writer's block. Finally, in the revision stage, AI can review the full work against a teacher-provided rubric.

It could work similarly in mathematics. In the incubation stage, AI can help by reframing a problem or reminding the student of the information they know and the mathematics concepts available to them. During experimentation, AI can show how different approaches might be effective for solving the same problem. As the student engages in perspiration by working through the steps of the problem, AI can mirror their thinking back to them and pose questions to help them address mistakes or push themselves further. For revision, AI can work as a tutor, pushing students to understand the source of their mistakes and nudging them toward productive solutions without giving them away.

These approaches could apply equally to foreign-language classes, lab work, art and design projects, and most any other understanding or work students develop, such as group presentations, music practice, research projects, and so on. Each of our learning contexts now has the opportunity for ongoing feedback throughout the learning process.

While ongoing feedback is now possible, teachers might nonetheless set the pace of feedback in an assignment, asking students to work without support in some stages and then to turn to AI for feedback in other stages. An English language arts teacher, for example, might want their students to research on their own (incubation) and develop and outline their ideas on their own (experimentation), thereby showing their original thinking, but they might encourage students to leverage AI for help with the writing (perspiration and revision). Or a science teacher might invite students to research a subject with AI (incubation) but then design and execute an experiment on their own (experimentation and perspiration), thereby showing their mastery of the scientific method, before using AI to help with final edits (revision).

The purpose of most feedback is to produce a change; it is a call to action. If feedback comes at the end of a unit and there isn't a next action, then there isn't any opportunity for change (Dickson & Housiaux, n.d.). Now, instant feedback from an AI assistant, available at any hour, enables more opportunities for students to change. The timely availability of feedback makes the case for not only permitting but requiring progressive deadlines or other opportunities for iterative feedback throughout the creative process. In this way, feedback is not a question of *whether* but of *when*.

Clarity

In Grant Wiggins's (1993) work *Assessing Student Performance*, a student says to their teacher, "By the way, you kept writing one word over and over on my papers, but I don't know what the word means." The teacher asks the student, "What's the word?" and the student replies, "Vagoo."

Wiggins's (1993) story in which a student misreads *vague* as "vagoo" can be misunderstood as a joke about student misunderstanding. It is certainly designed to be funny. But at a deeper level, Wiggins, as a researcher and writer, intends this story about an earnest teacher, an earnest student, and a funny moment of incomprehension to be not only a bit of humor but also evidence of a deeper problem: The teacher gave feedback that the student didn't understand, and this misunderstanding persisted "over and over." The vagueness the teacher sought to point out in the student's work might equally have described their own. The student's writing might have been vague, but so was the teacher's feedback.

To be effective, feedback must be descriptive and succinct. It must home in on the most relevant information. It must help students close the gap between what they intended and what they accomplished, and this requires intentionally focused thinking and writing by teachers, over and over again, for each student's work. Producing timely and relevant feedback like this in consistently clear language requires endurance. It takes enormous skill, patience, and discipline.

And it is the perfect assignment for a machine.

When forty-three students took part in a study in which they received feedback on their writing from both ChatGPT and a human tutor and were asked their preference, the results were remarkably split: "About half the students preferred receiving feedback from a human tutor, and half preferred AI-generated feedback," write the authors of the study (Escalante et al., 2023, p. 13). They continue:

> Those that preferred sitting down and discussing their feedback with a tutor cited the face-to-face interaction as having affective benefits, such as increasing engagement, as well as benefits

> for developing their speaking abilities. Those that preferred AI-generated feedback primarily cited the clarity and specificity of the feedback as being useful for improving their writing. (Escalante et al., 2023, p. 13)

These results are consistent with other findings that AI-generated feedback is often both clearer and more detailed than human-generated feedback (Almegren, Mahdi, Hazaea, Ali, & Almegren, 2024; Dai et al., 2023; Kurt & Kurt, 2024; Li et al., 2024; Wu et al., 2025).

"Please forgive the long letter," starts the line variably attributed to Mark Twain, Cicero, Blaise Pascal, and others. "I didn't have time to write a short one." It takes time to see clearly and write succinctly. And time is one of teachers' scarcest commodities. Clear, relevant, and timely feedback from AI enables teachers to focus their scarce time on the parts of the student learning experience that matter most.

In practice, this means that feedback might follow a Claim-Evidence-Reasoning structure. The *claim* is the degree to which a student achieved or fell short of achieving the learning objective. The *evidence* for this claim comes from explicit references or quotes from the students' work. Then the *reasoning* explains why this excerpt of student work does or does not achieve the learning objective. The degree to which evidence and reasoning portions of the feedback statement are made explicit is the degree to which the feedback is clear.

Sometimes, teachers may choose to omit evidence and reasoning. This gives students practice seeing their work through the lens of the claim about their work, which is ultimately necessary for them to internalize the learning and adjust their own actions in the future. However, this should be an intentional learning design choice, not an oversight.

Clear feedback is specific, tangible, and observable. The taste of a soup is immediately observable to a chef. If a chef had to learn how to cook based solely on people's reactions—without ever tasting the soup for themselves—the learning process would be much more challenging. For students, the Claim-Evidence-Reasoning structure observably shows where and why changes need to happen. Without this, students' confusion about how claims in feedback relate to their work introduces a break in the feedback cycle and undercuts their ability to apply the feedback to their actions.

To provide clear AI feedback for students, try the following.

- **Ask for evidence:** Just as we require students to show their work to help them clarify their thinking, the same expectation applies to AI. Prompting AI to cite specific evidence from student work to justify its judgments can help make sure that it sticks to the reality of what the student wrote, and it has the added benefit of providing an audit trail that can be manually inspected.
- **Provide examples:** Showing AI an example of how you would provide feedback will help it compose its own in the way you prefer. Give it a particular learning objective and real student work, then model what you expect. This has the added benefit of showing AI your tone, style, and voice, which it can try to emulate as well.
- **Empower students to seek AI feedback (safely):** While working on a project or assignment, set aside a class period to show students how to elicit feedback from AI. Demonstrate how to upload the assignment and rubric, and then critically assess the feedback from AI. What was helpful? What wasn't? Why or why not? In addition to helping them with the assignment by giving them personalized feedback, these steps also offer them the opportunity to build their metacognition about types of feedback as well as their AI-literacy skills.

Crucially, clear feedback does not explicitly tell students exactly what to do next. This would rob them of the chance to close the loop for themselves and would impinge on the very agency we are trying to foster.

Relevance, Timeliness, and Clarity and Student Agency

Knowing the goals of an assignment and exactly which actions led to success or failure to achieve those goals closes the feedback loop and allows for learning. But the exact wording for maximal agency is hard to craft. Feedback is full of balancing acts. Vague comments on writing don't help a student know how to proceed. On the other hand, detailed directions that provide or lead students straight to a solution don't help students master skills either; they create dependency. So, teachers should provide feedback that mirrors what they see students doing and then poses questions or frames next steps—not directing them what to do but opening up avenues for them to explore.

Another similar balancing act is required when setting the scope of feedback. Encouraging students to achieve easy goals, for example, doesn't lead to growth as much as keeping students focused on challenging goals. But, at the same time, feedback about larger, more complex, and more challenging work is less effective than feedback on smaller steps within complex tasks (Kluger & DeNisi, 1996; Shute, 2008). So, teachers provide feedback on small steps while referencing large, complex goals. At its best, actionable feedback balances these various needs; it builds on students' successes while showing areas for growth, keeps sight of big goals while focusing attention on details, and points to relevant next steps without doing the work for students.

Not only is this possible with AI, but it is already happening to some degree. AI tools now provide live feedback for students doing problems on interactive whiteboards. Students narrate their thinking while writing out their work. AI then analyzes both what they say and what they write and responds to their thinking, offering guidance on how to proceed. Students can then incorporate this feedback and retry the problem, finally submitting whichever iteration of their work they feel proud of to the teacher. Further, these AI tools then aggregate information from conversations with each student to provide high-level overviews for the teacher of the kinds of misconceptions that are prevalent in the class, informing how all-class time might be used most effectively to support the greatest number of students. Snorkl (https://snorkl.app) is one such tool.

Students can now receive feedback like chefs: They can taste the soup as many times as they want, being intentional about what they are tasting for, with plenty of time to make adjustments. Then they can close the feedback loop and learn from their actions. This sets them up not only for greater understanding but for more agency during and after learning.

Teachers no longer have to provide individual feedback on every iteration of student work, but that doesn't mean their jobs are strictly easier.

AI in Practice: Classroom Example

Ten years ago, when Mr. Mitchell was just beginning his teaching career, he often found himself in his classroom at 2 a.m. It would be late in the semester, and he'd be grading stacks of history essays and writing comments for his students. Another young colleague in his department would be down the hall doing the same thing. Coffee and soda kept them humming along, but it would be a tough weekend when the sleep deprivation caught up with them. He remembers being exhausted often and needing a nap every day when he got home.

Now, it's different. The students in his U.S. history class are working on an essay on the provocative question of whether the First Amendment in the Bill of Rights should apply to students in school. He has told students that they should come up with their ideas on their own, and they should write their own first draft, but then they can use AI for feedback on revising and improving their writing.

Mr. Mitchell has provided a link to an AI feedback tool in the learning management system. He has given instructions for AI to review student work against a rubric that focuses on four structural elements of the essay: (1) stating a clear argument, (2) ensuring each paragraph serves the thesis, (3) assessing whether transitions between paragraphs are logical, and (4) determining whether quotations from the Bill of Rights are properly integrated into the writing. He has also clearly indicated both to students and to AI that he will be assessing students' work based on whether they made connections between the principles in the First Amendment and the details of their individual school community. Specifically, he'd like students to discuss (1) where public speech happens at their particular school, (2) how legal limits on freedom of speech are set in public spaces outside of school, and (3) whether their school, which is full of students (minors), should have the same rules as public spaces for adults. He notes that AI is prepared to give feedback on the structural components of the essay, but students must draw on their own experiences at their school to succeed in the essay.

While the students are writing their drafts, they copy and paste their writing into AI and receive as many rounds of feedback as they wish while they make revisions. Mr. Mitchell receives an overview from AI that summarizes how the class is engaging with the tool. How much guidance did the class as a whole ask for on the structure and form of their writing? What was the difference in strength from first draft to submitted draft? Were the students equally strong in making a clear claim as they were in ensuring their paragraphs supported their thesis? Did they require multiple changes to integrate quotes properly?

The overview suggests that most students were able to reach appropriate levels of mastery after a first or second round of feedback, except for when integrating quotations. When quoting the text, nearly all students required several rounds of correction. Additionally, several students required significant guidance across all categories. Some students didn't use AI at all.

With the overview in mind, Mr. Mitchell makes a note to review quote integration in an upcoming class and then turns his attention to each final paper. Each student's submitted writing includes a two-to-three-sentence summary at the beginning, generated by AI, describing the edits each student made with support from AI. Mr. Mitchell reads each note, then continues to the essay. Across the board, the syntax and structure of the students' work are strong; they received AI guidance on these areas. When he reads, he is looking for the substance in the students' arguments and how each student made connections between the First Amendment and their school context. He looks for the criteria he mentioned, examples of which AI highlighted and annotated, ready for him to check and edit as needed. He doesn't have to stop to correct grammar, specify that paragraphs need transitions, or explain any of the three main ways to integrate quotes. He dives into the substance and leaves comments on the connections the students made between what they are studying and their lived experiences.

Several days later, when the class engages in discussion about this work, Mr. Mitchell plans both to assess oral mastery of the ideas in their writing and to deepen their understanding through a discussion with their peers. He has seen some adjacent and opposing ideas in the students' writing, including two students who argued very strongly on opposite sides of the question. One student advocated for total freedom to speak, even about offensive ideas, because those needed to be tested and rejected socially. Another student suggested that kids aren't mature enough yet to handle the responsibility of totally free speech. This will be a good starting point for class dialogue. He'll ask them to present their perspectives, and the discussion will fill in the space in between.

By the end of the unit, the students will have received multiple rounds of feedback on their writing, from both AI and the teacher. They'll also have submitted their work with pride, knowing that it is strong in both its mechanics and its form. Mr. Mitchell will have been able to spend his time outside of class on a few key tasks: guiding AI on what feedback to give, reading student work for its substance (with less time focused on mechanics), and shaping class time to address identified student needs.

Teachers as Executives

Teachers' primary relationship is with their students. But now, they also have a team of assistants. What once was the province of college professors, with their teams of teaching assistants, is now the experience of all teachers. We all now have a team of assistants, and these assistants require our executive management.

The first time that employees take on a leadership role with another person reporting to them, they often learn that two people working together do not complete the work of two people; they might complete the work of 1.8 people. Why is this? Part of the work of being a supervisor is supervising: coaching the new team member, setting objectives and guidelines, collaborating on schedules, evaluating work, and more. This takes time away from individual productivity, but it also ensures that the team's collaborative work is in alignment and leads to greater outcomes. A supervisor may be able to produce less of their own work, but the team collectively produces more—and, ideally, higher quality—work than each member could ever produce alone.

This is similar to how we work with AI. If AI is like an assistant, then part of our time is spent coaching the assistant, setting its objectives, evaluating its output, giving it feedback, and more. The result is that students receive more total feedback, and teachers spend less time providing certain types of feedback, with some of that time instead spent managing AI.

What does this look like in practice? In the first-pass feedback model we described in this chapter, students experience initial rounds of learning and feedback with AI, then gather further feedback from their teachers. Notably, the teachers will have already seen the feedback their AI assistants previously provided, just like doctors see patients after having reviewed the notes of nurses who worked with the patients first. This structure enables students to gain more opportunities for learning and teachers to gain more insights into their students.

In this way, teachers are now acting as executives. They still work primarily with students, but they also now lead teams of assistants that they direct and guide in order for them to be effective. What new skills will be required of teachers?

In many ways, the skills of executives and teachers already overlap.

More than a century of leadership study has articulated a growing vocabulary of executive skills (Drucker, 2008; Kotter, 1996; Mintzberg, 1973). These skills go by different names but often include similar elements: setting a vision, defining and delegating work, motivating a team, measuring progress and providing feedback, integrating multiple perspectives, choosing a course of action, and adjusting and adapting.

These may sound familiar to teachers because they already are managers: of students. Setting a vision is like establishing learning objectives, defining and delegating work is like designing and assigning student work, and so on. Managing AI feedback assistants involves similar skills: Choosing a course of action is like deciding which recommendations from an AI assistant to execute, and adjusting and adapting how your team works is like revising and improving how you use your AI assistant from year to year. Table 3.1 shows some of the skills involved in managing AI feedback assistants and how they align with working with employees and students.

Having a team of AI assistants doesn't mean that teachers are not reviewing their students' work; it means that when student work reaches them, it will have had the benefit of feedback and iteration. This allows teachers to focus more on its substance and on student learning, rather than on mechanics. Their relationships to the students and their work are not decentered by but rather reinforced with the use of AI.

Bias, Equity, and Trust in Feedback and Grading

Should machines even be giving feedback in the first place? Should they be assessing how effectively students completed assignments or be recommending grades? Isn't this kind of evaluative work rife with risks of bias? If AI tools are trained on the internet, then doesn't AI carry with it all the biases of its training set?

Evidence of bias in AI is well documented (Blodgett, Barocas, Daumé, & Wallach, 2020; Bolukbasi, Chang, Zou, Saligrama, & Kalai, 2016; Buolamwini & Gebru, 2018), but so is evidence of bias in people (Greenwald & Banaji, 1995; Greenwald, McGhee, & Schwartz, 1998; Kang, DeCelles, Tilcsik, & Jun, 2016; Thorndike, 1920; Wason, 1960).

The alternative to using biased AI isn't a bias-free classroom—it's bias-prone people. At least with the technology, we might more predictably understand its biases and address them. We highlight AI's shortcomings here not to say AI shouldn't be used but rather to ensure that an understanding of bias is part of the discussion of AI's role in schools.

***Table 3.1:** Managing AI Feedback Assistants*

Skill	Working With Employees	Working With Students	Working With AI Feedback Assistants
Setting a Vision	Frame the problem, name the stakes, and draw the moral boundaries within which the work must proceed.	Design learning experiences based on learning objectives.	Explain the assignment, its purpose, its context, students' characteristics, and any other relevant framing.
Defining and Delegating Work	Charge the team to draft, simulate, or forecast—always with a clear brief and a plan for how the output will be judged.	Write student instructions and assign the work with clear milestones.	Clarify the specific elements of student work that should receive feedback and the specific types of feedback desired. Include a rubric and examples.
Motivating a Team	Create a positive, supportive, and productive workplace culture.	Motivate students with an emotional hook (a personal or social connection) and create a positive classroom culture.	For now, this skill is the least applicable. However, once AI agents take on a more active role, it may be important to coordinate their communication among each other.
Measuring Progress and Providing Feedback	Look for what is missing, challenge thinking, and ask for specific improvements.	Provide relevant, timely, actionable, and clear feedback for students.	Provide feedback for AI to tweak its tone, voice, approach, and the specific ways it applies the rubric to the students' work.
Integrating Multiple Perspectives	Weave the team's findings with human stakeholder voices and the contextual nuance that no individual fully captures.	Combine all the students' contributions into class learnings.	Triangulate AI feedback with your own assessment of the student work, as well as with AI feedback on other assignments.
Choosing a Course of Action	Weigh trade-offs, take ownership, accept responsibility for consequences, and communicate the rationale.	Make decisions about what to teach next and how, informed by what was just learned.	Decide on a summative grade based on the AI assistant's formative feedback and preliminary grade suggestions.
Adjusting and Adapting	Treat each creative production cycle as a feedback loop, refining both the questions and criteria.	Improve each assignment each year based on experience.	Improve the feedback ecosystem for each unit from year to year so it is more supportive of both students and teachers.

While bias in technology—as in people—can never be fully removed, it can be identified, reduced to the smallest degree possible, and made transparent. This mitigation must be a key responsibility for technology companies as well as individuals. At the same time, AI literacy in schools must include skills to recognize and counteract bias in AI. The coordinated efforts of technology companies, educational institutions, and individuals can yield the most productive and least harmful contributions to society.

Social and pragmatic perspectives may align on this. Given the prevalence of AI in high-earning professions, providing AI access to the least advantaged students may be a *necessary* step for promoting a more equitable society. Restricting disadvantaged students from using AI because they might be vulnerable to bias from the technology may not support them; on the contrary, it may limit access to and understanding of a powerful tool emerging in the modern economy. As educators, we should be facilitating student literacy with AI (see chapter 5) and how to apply it to academic and professional settings. We should enable deep, knowledgeable understanding of what AI can do, what its risks are, and how students can leverage it to increase their own agency (Bozkurt et al., 2024).

The same pattern is playing out on a global scale. The greatest excitement for and adoption of AI has consistently been in emerging markets—rapidly developing countries with growing infrastructure and strong incentives to leverage new technologies for economic and social development (Google & Ipsos, 2025; Ipsos, 2022, 2023; Johnston et al., 2024). These countries have the infrastructure to leverage AI for their professional benefit, and with AI, they can increase opportunities for social mobility. If our objective is a more equitable society, we may want to follow and support this adoption and make this new tool as widely accessible as possible.

Student Perceptions of Bias

Further, discussions of bias and AI have an entirely different angle that is often overlooked: how students *perceive* bias in feedback. As much as we do our best as teachers to support our students without bias for or against them, our students may yet hold perceptions about what we believe. Students may believe, for example, that their strong marks are because their teachers like them. Or, conversely, every teacher has heard a variation on a student saying that they received a bad grade because their teacher *doesn't* like them. This bias that students falsely (or accurately) perceive in teachers is often an obstacle to learning.

But this appears to be different with AI. Some student surveys suggest that when students receive feedback from both teachers and AI, they perceive AI to be *fairer* than their human teachers (Chai, Ma, Wang, Zhu, & Han, 2024). It's notable that when researchers controlled results so that both humans and AI included explanations for their evaluations, gaps in student perceptions of fairness decreased, but students still perceived AI as having greater information transparency. These student perceptions of fairness amount to feelings of trust, and teachers will recognize that when a student trusts a teacher, the student is more open to feedback. This early research suggests that students extend this trust of fairness more readily to AI than to people (Ruwe & Mayweg-Paus, 2024). This means both that we need to continue earning our students' trust and that we have an opportunity as teachers to leverage AI for student feedback—because students seem to trust it more.

If students perceive AI feedback to be fairer than feedback from their teachers, then leveraging AI assistants for feedback may give students greater confidence in how they proceed with their work.

The Irreplaceable Component of Feedback: The Human-AI Loop

AI with no feelings don't deserve to teach. It's so obvious.

—ADAPTED FROM THE LYRICS OF THE *KPOP DEMON HUNTERS* SONG "TAKEDOWN"

When the U.S. Department of Education's Office of Educational Technology released its initial guidance on the use of generative AI in teaching and learning, its first recommendation was "Emphasize Humans in the Loop" (Cardona, Rodríguez, & Ishmael, 2023, p. 53). The report expands on this recommendation as follows:

> We envision a technology-enhanced future more like an electric bike and less like robot vacuums. On an electric bike, the human is fully aware and fully in control, but their burden is less, and their effort is multiplied by a complementary technological enhancement. Robot vacuums do their job, freeing the human from involvement or oversight. (Cardona et al., 2023, p. 53)

This is sensible guidance. Notably, however, just one year prior—and less than two months before the release of ChatGPT—the Stanford Institute for Human-Centered AI proposed different guidance: "AI in the Loop" (Lynch, 2022). The principle, it turns out, is the same, but the framing is reversed. The language from the U.S. Department of Education report insists that humans be a part of any AI feedback cycle. The language from Stanford, on the other hand, insists that we think of the loop as a fundamentally human cycle that AI can be inserted into, in order to enhance what humans do.

The two framings establish different foundational contexts. In both framings, however, humans and AI work together in a loop to make decisions because humans have at least three roles that are irreplaceable by AI.

Human Sensitivity

First, humans are more sensitive to human context than machines. At least for now.

Pragmatically, technology is increasingly aware of the contexts of our lives. Already, we carry digital devices around with us nearly everywhere we go. They measure our heart rates, and they track every movement we make, not just when we travel by car or by foot but when we sit, stand, or shift our weight. They even hold some of our most intimate communications. The technology we use now knows more about us than most of us are aware. Given this awareness of our lives, the pragmatist might argue it has the potential to provide increasingly useful feedback in educational contexts.

And yet, what happens in a classroom cannot easily be captured by even this sophistication. Personal lives, family lives, school contexts, social and political events—these have an impact on student learning.

Humans read these contexts intuitively, and, crucially, we discern the relevance of how they affect our classrooms. Like our students, we educators have our own lived experiences that lead us to understand and empathize with the lived experiences of our students in ways that AI—at least for the foreseeable future—cannot.

When it comes to making decisions about grades, having a human in the loop ensures we can appropriately take into consideration the context of our students' experiences.

Ownership and Accountability

Second, humans in the loop are essential for maintaining ownership and accountability, both professional and ethical. Professional ownership and accountability ensure alignment with learning goals. They ensure that assessments and grades reflect not only standards and objectives but also the teacher's informed understanding of students' progress, needs, and potential. Ethical accountability and ownership mean that feedback is also morally sound. When systems automate feedback, biases like those we described earlier in this chapter might go unchallenged, and patterns of error might scale. Human oversight ensures that if feedback or grading schemes become harmful, someone is involved who has the authority and moral responsibility to intervene and correct the course.

Appreciation and Care

Last—and perhaps most powerfully, from a humanist perspective—a human in the loop ensures that if a student works hard on something, they can know that their hard work is seen by a person, not just processed by a machine. Students want to be cared for, not just taught (Gallup & Walton Family Foundation, 2024). Human oversight and feedback communicate more than information—they convey that someone is paying attention. A student who puts effort into their work should know that a human recognizes that effort.

Teachers must be the final arbiter of what feedback and grades look like. Their human judgment provides insight into the experiences of students. Their professional accountability ensures feedback and grades are not only aligned but also ethical. And their humanity means that students know that their work is seen and recognized. While grades may be informed and even recommended by layers of feedback from machines, they must ultimately be the domain of people.

Review and Next Steps

We have all seen them on the side of the road, affixed to a telephone pole or parked on a towable trailer: the blinking sign displaying both the speed limit and the speed of our car. The feedback is relevant, timely, clear, and thus actionable. And it is one of the most effective interventions for changing the behavior of drivers (Gehlert, Schulze, & Schlag, 2012; Williamson, Fries, & Zhou, 2016).

The arrival of AI provides an opportunity for students to gather unlimited first-pass feedback. And with this feedback source for students, teachers' relationship with students changes, as does their workflow. Whereas once teachers were the primary source of feedback for students, now, they are one of several sources. Whereas once teachers read every submitted iteration of student work, limiting how much feedback they could provide, now, they manage AI assistants that provide more feedback than teachers ever could.

Machines have twenty-four-hour availability and a near infinite capacity for knowledge, but they still lack a rich contextual awareness of the life and work of a student. Technologists should not pretend that a machine can meet all of a human's educational needs, but they should continue to explore what parts of our human experience benefit from focused support. At present, students trust AI to be neutral, clear, and consistent, but they still trust humans more to understand affective experience (Escalante et al., 2023; Ruwe & Mayweg-Paus, 2024).

Of all the advancements that generative AI has brought us, AI-powered feedback for students may be the most significant opportunity for learning. Feedback is central to growth, and now—with the appropriate structures—feedback is continuously available.

To give an analogy, evolutionary biologists suggest that life evolves very slowly for long periods of time, but then it goes through brief periods of rapid change, often due to sudden shifts in environment. This theory is called *punctuated equilibrium*. Similarly, AI has suddenly shifted the environment of education, and in our moment of punctuated equilibrium, continuous feedback may prove to be the largest, most positive change for learners—if we collectively get it right.

At present, students know their teachers are always there to see their work. In this new era, with a human–AI loop, they can be confident that their teachers will still be there. Now, however, they can also know that they have an additional source of feedback to support them. They now have a personal assistant, too—one that can help them understand and one that can help them take what they learn and create with it. AI assistants for students will be the focus of part 2.

Before we move on, here are some potential next steps for educators, school leaders, and those working in the technology industry.

For Educators

The following are some considerations for educators related to AI for classroom feedback.

Experiment with AI-generated feedback on student work. Remove personal identifying information from a student's work, upload the work to an AI tool, and ask AI for feedback. How helpful was the feedback? Try submitting additional assets, like the assignment, a rubric, or a sample of model work. Try sharing your learning or course objectives, and ask for feedback that is specifically related to those objectives, tying to excerpts of student work as evidence. Ask it to provide reasoning. How does this change the feedback? Does it give too much or too little? Is it actionable? See what works and what doesn't.

Facilitate AI-generated feedback for your students. Within the context of approved AI tools at your school, build AI-provided feedback into your students' learning experiences. Which steps in your students' creative process might benefit from feedback? Which sections do you want to be the sole work of the students? See part 2 for more insight on how to approach AI feedback and integration into student work.

Look at AI-generated feedback with students. To advance both their and your AI literacy, workshop a report, project, assignment, or tool together with students in a class. Discuss the work, model for the students what asking for feedback looks like, and, crucially, evaluate the feedback with the students. In this experience, share the assignment, any available rubrics or criteria, and other relevant resources. Further, model what back and forth dialogue with AI looks like, asking for further explanation and even challenging the feedback.

For School Leaders

The following are some considerations for school leaders related to AI for classroom feedback.

Ensure equitable AI access for students. For AI feedback to serve as a tool for greater equity, school leaders must ensure that all students have consistent access to the devices, bandwidth, and approved tools that support its use. This includes both access during school hours and thoughtful planning for access beyond the classroom, especially for students who may not have an internet connection at home. This planning includes clear communication to parents about the role and importance of AI in student learning. Without equitable access, the benefits of AI feedback will reinforce existing divides instead of helping to close them.

Integrate AI literacy into faculty and student development. Develop ongoing professional development opportunities that help teachers understand how to guide, critique, and contextualize AI feedback and how to teach students to do the same. For students, experimentation should include not just how to use the tools but how to recognize when AI feedback is useful, when they need to challenge it, and how it connects to their broader learning goals. Embedding these learnings into advisory programs, student orientations, and classroom routines will ensure that students develop healthy, empowered relationships with AI. See more on AI literacy in chapter 5.

For Technologists

The following are some considerations for those working in the technology industry related to AI for classroom feedback.

Build feedback tools that are customizable for a variety of educational objectives. Sometimes, teachers want students to brainstorm with AI; other times, they want students to generate ideas entirely on their own. Whether you are developing a general chatbot for educational purposes, in which all functions take place in a central chat bar, or whether you are developing a Swiss Army Knife–type AI tool with many customized functions, ensure that teachers can easily modify the way students use it to focus on any of the different stages of student work.

Visualize feedback summaries for teachers. One of the most powerful ways AI can support teachers is by helping them see patterns across their classrooms. Technologists should build tools that aggregate student interactions with feedback systems and display them in dashboards that highlight common strengths, common misconceptions, and gaps in usage. These summaries should be easy to interpret and actionable—for example, showing that most students struggled with transitions between paragraphs or that very few students requested help with evidence integration. Equipped with this kind of visibility, teachers can better tailor their instruction and adjust their pacing.

Incorporate high-information feedback scaffolds. Move beyond binary or corrective feedback, and design interfaces that support high-information feedback, including tools that structure responses into clearly labeled sections with clear, actionable steps for students to take. This might look like a table breaking down grammatical errors by type, giving an explanation, and providing possible next steps or a side panel showing how a student's paragraph aligns with rubric criteria. Ideally, teachers can edit or annotate these scaffolds to reflect classroom language and reinforce learning routines already in place.

Part 2

AI FOR STUDENTS

The one who does the work does the learning.

—TERRY DOYLE

If students learn by doing, what will they *learn* if they *do* with AI?

This question speaks to the essence of both our hopes and fears about AI. Will students use AI as a tool to enhance their thinking, thereby learning to stretch themselves, explore, experiment, and grow in ways we are only just discovering? Or will they use AI to replace their thinking, thereby learning to depend on AI and failing to develop their own skills?

While everything we've explored so far examines how teachers can use AI to advance their objectives, none of it matters if students don't use their technological resources responsibly. And whether they use resources responsibly depends not only on the lessons we design but also on students' motivation, on the matching of competency with difficulty, and on their feelings of agency.

We now teach in a time when every student has not only their own AI tutor but also their own creative team. With motivation, engagement, and agency, they can learn most anything and create most anything, all on their own.

Or can they?

Part 2 explores how students can and do leverage AI outside of class.

- **Chapter 4, "The Learning Assistant":** In chapter 4, we look at the work of what has historically been imagined as AI tutors. We ask how AI can help students access learning, but we position AI in the context of a long history of failures at machine-driven learning experiences. We also explore how what is emerging today may show signs of a new trajectory. How can machines contribute to student learning outside the classroom? Can they navigate our messy human experience of learning and our richly intertwined experience of thought, feeling, and embodied sensemaking? What might they need to be able to do? What can they do already?
- **Chapter 5, "The Doing Assistant":** In chapter 5, we shift from student understanding (learning) to how students apply and demonstrate their understanding (doing). What does it mean that every student now has a creative team to help them write, present, draw, and design? In this age of AI, what skills will students need to manage their creative teams? How can we design learning experiences so AI supports student work instead of replacing it?

The vision of always available, infinitely knowledgeable AI tutors is a powerful one. The possibility of providing support for every student at any time in whatever mode they need is inspiring. But it is only a useful vision if it recognizes the messiness and complexity of the human experience. In the two chapters ahead, we explore the challenges and risks inherent in this work, but we also suggest that this vision is newly within reach.

So, what will students learn? They will learn whatever they *do*. And our role as educators is to design experiences and set the conditions for our students to be empowered by their AI assistants to do work that develops the skills and dispositions they need to succeed in a new world.

CHAPTER 4

The Learning Assistant

In Pursuit of the All-Knowing, Endlessly Patient AI Tutor

Memory is the residue of thought.

—DANIEL T. WILLINGHAM

The history of educational technology is littered with the husks of AI tutors, their innards hollowed out by broken promises to revolutionize and personalize learning. These teaching machines revive like zombies, fed by the new technology of each era, intoning again and again, "This time, we will automate tutoring for every student. This time, it will be different." Until, like the zombies before them, they, too, heave and collapse, leaving their carapaces hanging on walls of classrooms, languishing in warehouses, or boxed in the corners of researchers' labs—only to be revived anew, lurching back into and then out of educators' lives again and again with the emergence and obsolescence of each new technology.

If this reads like a horror story, it's because for some educators, the arrival of a new technology is a source of dread. This is especially true for teachers who have put faith in technologies that ultimately added layers of complication to their lives without substantially improving outcomes or workflows.

Some see the AI boom as the nadir of another kind of zombification: a digital zombification in which schools now gather platters of data on student performance to support the insatiable appetites of edtech data systems that ultimately reduce students and faculty to machine-feeding button pushers and data points.

"To what end?" many educators ask. "Have we brought significantly more students up to baseline?" Some exceptional stories do emerge, and these speak to the real opportunities that technology enables. But the data shows that often, sometimes 95 percent of the time or more, students disengage from their tech tools and fail to complete what they started online—never mind having achieved meaningful gains (Holt, 2024; Reich & Ruipérez-Valiente, 2019).

This does not invalidate the often noble efforts of edtech developers, whose energy has established an ecosystem committed to supporting teachers and learners. The edtech ecosystem has yielded numerous tools that have driven student progress in a variety of contexts. But this zombie horror narrative does assert clearly that we have yet to realize the full aspirations that educational technologists have for facilitating progress at scale through individualized AI tutors.

The First-Mile, Last-Mile Problem

Mathematics educator Dan Meyer succinctly describes the source of the issue with a reference to the first-mile, last-mile problem (Meyer, 2024). In a city, public transportation moves people from one place to another place. But the challenge in transit is often getting people from their homes to the first stop (the first mile) and then from their final stop to their destination (the last mile). A transit system might work beautifully, but if people can't get in and out of the system effectively, then it ceases to serve its purpose.

So it is with educational technology. In schools, technology similarly helps move students from one place to another. But this movement is metaphorical; educational technology helps students move from one level of mastery to another. In the context of an AI learning assistant, then, the first mile might be whether the student engages with the technology to begin with. Do they have access to the technology? Can they use it easily? Do they know what to do when they don't know what to do?

When technology is used in the classroom, a teacher shepherds students through the learning experience and closes the first mile by guiding them to and through the experience. But does tech-enabled learning work when students use it alone at home? This is the vision, but if the students aren't engaged in the technology intellectually, emotionally, or otherwise, then learning doesn't happen.

Stated more starkly, in public transportation, if you never make it to the bus stop, you miss the bus and do not reach your destination. It is similar for students; if a student is confused by an app, they put it down, and learning doesn't happen. If they don't have the prior knowledge to understand an online video, or if they get frustrated by a mathematics problem online, or if they are distracted by the other things going on in their life and can't get themselves to care, learning doesn't happen. This isn't the fault of the student or even the technology—it's a simple fact of our complex human lives.

We just don't plug into learning experiences and learn like machines. We don't have operating systems with formulaic inputs and outputs. We learn like complex, emotion-rich human beings. Teachers know this from on-the-ground work every day, which is why they speak so passionately about the importance of *engagement*. Learning is not a simple mathematical function with predictable independent and dependent variables. Instead, learning is relational, full of social exchange, and unfailingly tethered to our emotional lives and our many sources of motivation. We are messy and unsure, we lack perfect self-awareness, and we usually require encouragement and companionship to succeed.

The zombie vision of AI tutors has produced, for decades, teaching machines that have treated people not like humans but like machines. And the result has been a string of failed enterprises.

A New Hope

Despite this grim introduction, we believe that AI-driven technology is approaching a threshold, if not yet crossing it. Despite the failed history of teaching machines, we believe that this time—and we understand the irony of saying this—it is different. To be clear, machines will never replace teachers, and they will not be fully successful AI tutors for all students, but the capacity for machines to provide supplemental learning experiences for students has expanded.

This chapter starts so cynically because it would be folly to write about AI tutors without acknowledging and learning from the undeniable failure of previous attempts to drive systemic change with AI tutors.

So why might this time be different?

The changes brought by AI are manifold, but the most significant is AI's increased sensitivity to human capacities—not only our cognitive or intellectual capacities but also, and especially, our emotional and physical capacities. Machine learning and generative AI have accelerated technology's ability to engage with us in more human modes. It starts with chat, but it extends beyond text-based interaction, both in how it presents itself and how it reads us. Voice chats and image and video generation now simulate human voices and visages in ways that we perceive as lifelike (such as HeyGen [www.heygen.com] or Midjourney [www.midjourney.com]). Further, cameras read our facial expressions, microphones discern feelings in our voices, and data models detect patterns in our behaviors. This awareness of and interaction with human behaviors enable today's technologies to recognize human emotions in ways that might better support teaching and learning. Technology won't replace teachers and tutors, but it is better poised to support more of the complex work that has historically been a solely human domain.

For some, this is a terrifying prospect, one in which machines are trespassing on hallowed human grounds. There are risks of manipulation, of privacy violations, of emotional harm, of the replacement of agency with algorithms, and more. These are enormous risks and must be carefully accounted for by technology developers. At the same time, for teachers, this new ability for technology to engage both in a more human manner and in a way that reads more of our human experience creates opportunities to more meaningfully understand and engage with students.

This is a pivotal moment in the history of our human relationship with technology. It is fraught with risk, but it is also full of opportunity.

Commercial interests in this opportunity will inevitably move quickly. Those with educational interests must move equally quickly.

Head, Heart, and Hands: Foundations for Learning

Historically, most educational technology has been built around cognitive learning principles. Adaptive learning platforms have dynamically tailored the difficulty level of content to student performance, providing either easier or more challenging material depending on how a student performs. This would seem to align with Benjamin S. Bloom's (1956) taxonomy of educational objectives, one of the stalwart guides educators use to assess the quality of learning experiences. Often used as a shorthand hierarchy for cognitive complexity, Bloom's ordering of educational objectives begins with basic *knowledge* and *comprehension*, rises through *application* and *synthesis* of knowledge, and peaks with *evaluation* of knowledge. Later revisions of the taxonomy removed *synthesis* and placed *creativity* at the top of the taxonomy (Krathwohl, 2002). If learning experiences followed good educational-design principles, the thinking has gone, then moving students intentionally toward the top of the taxonomy would seem to be a design priority for effective learning experiences.

What most people miss, however, is that Bloom's taxonomy was developed across three domains of educational objectives. We are most familiar with the thinking-based cognitive domain, as previously described. But Bloom followed his initial publication with a second taxonomy of educational objectives for the emotion-based affective domain of learning (Krathwohl, Bloom, & Masia, 1964). Shortly after, a third taxonomy addressed educational objectives for the action-based psychomotor domain of learning (Simpson, 1966). Table 4.1 (page 100) details these taxonomies.

Together, these three domains communicate that learning is not just one kind of experience; it is not just cognitive development. Instead, learning is an integrated experience that involves our physical bodies and the emotions that we experience through them (Immordino-Yang & Damasio, 2007).

Table 4.1: Bloom's Taxonomies of Educational Objectives

Cognitive (1956)	Affective (1964)	Psychomotor (1966)
Evaluation	Characterization	Origination
Synthesis	Organization	Adaptation
Analysis	Valuation	Complex overt response
Application	Response	Mechanism
Comprehension	Reception	Guided response
Knowledge		Set
		Perception

Other psychologists, researchers, and theorists, both before Bloom and since, similarly asserted that learning integrates head, heart, and hands. The Swiss developmental psychologist Jean Piaget (1951), the Italian physician and educator Maria Montessori (1967), the American philosopher and educator John Dewey (1938), and many others have written about integrated learning experiences for children, with many of these theories centering on experience and play. Educators often understand this intuitively; students learn best when they are emotionally and physically engaged, not just intellectually engaged. Even further, sometimes they *must be* emotionally and physically engaged *in order to be* intellectually engaged.

Learning involves thinking, feeling, and doing. But educational technology, for the most part, has been focused almost exclusively on thinking. Although thinking is undeniably crucial to learning, it is not enough, and this narrowness in focus been a significant part of its historic failures. Technologies intended for students' self-directed use must also acknowledge and engage the emotional and embodied experiences of learning, or else they almost inevitably fail in the first mile.

Today, with the advent of AI technology, tools are beginning to successfully acknowledge and engage students across the learning domains that Bloom and others have described.

This chapter explores how advances in AI can engage the head, heart, and hands in ways that open new opportunities for teachers and students. These experiences that combine cognitive, social-emotional, and embodied learning may contain the seeds that bring the long-awaited, often attempted, and frequently failed vision for an AI tutor—or AI learning assistant—to fruition.

Head: Early AI Tutors

Intelligent tutoring systems are not new. As early as the 1970s, systems used natural language interfaces to imitate aspects of human tutoring (Carbonell, 1970). Even in early iterations, software was designed to infer a student's capabilities and tailor instruction accordingly (Anderson et al., 1995; Barr et al., 1976). These early systems operated in narrow domains and were limited by the technology of their times, but they nonetheless marked a shift from programmed instruction to adaptive tutoring.

Since then, the field has grown steadily, though it has cycled through periods of hype and disillusionment (Alkhatlan & Kalita, 2019; Sleeman & Brown, 1982). But through it all, and up to the present, the aim of technologists and researchers has been consistent: to replicate the effectiveness of one-on-one tutoring at scale.

What distinguishes intelligent tutoring systems from other learning technologies is that they have three components: (1) a domain model (what's being taught), (2) a student model (what the learner currently knows and can do), and (3) a pedagogical model (how to best support learning). The domain model is a representation of the relevant curriculum, which, in the early systems, usually consisted of a set of logical rules, problem-solving steps, or concept maps. Next, the student model is an estimation of the learner's mastery of the domain based on how the student responds to prompts from the system. And last, the pedagogical model uses the information in the student model to decide what comes next for the student: problems, hints, questions, feedback, or other responses (Shute & Psotka, 1996; VanLehn, 2006). More sophisticated systems even attempted to simulate Socratic dialogue or scaffold problem solving step-by-step, sometimes even modeling student misconceptions (Graesser, McNamara, Louwerse, & Cai, 2004; Koedinger, Anderson, Hadley, & Mark, 1997).

In very limited contexts, these intelligent tutoring systems have worked. Meta-analyses from the 2010s show that intelligent tutoring systems can improve student learning, and they often both outperform traditional large-group instruction and help students reach mastery in shorter amounts of time (Kulik & Fletcher, 2016; Ma et al., 2014). They have even supported more complex reasoning and problem solving (Koedinger et al., 1997; Ma et al., 2014).

However, the same meta-analyses reveal that these successes have typically been in narrowly structured domains, like algebra or computer science, and they have required heavy implementation support from teachers. Intelligent tutoring systems without human support and scaffolding or in more open-ended subjects have shown smaller effects, reinforcing the importance of teacher presence rather than self-guided use. In other words, intelligent tutoring systems have not historically achieved the vision of stand-alone tutors, but they have provided support for teachers and students in scaffolded settings.

Now, however, with the emergence of generative AI, the field has taken enormous leaps in new directions. AI today is more flexible and conversationally fluent than previous learning systems, and the result is a much broader range of utility across subjects. Systems like Khanmigo (www.khanmigo.ai) offer guided practice with immediate adaptive feedback (Hess, 2025; Jiang & Jiang, 2024; Zhang et al., 2024). Meta-analyses suggest these systems can support critical thinking and conceptual mastery across subjects (Wang & Fan, 2025).

But there is another important caveat. Whereas intelligent tutoring systems have been designed around structured, factual models of the curriculum and with clear rules for how to engage it, the language models underneath generative AI tools have no such grounding in reliable structure or fact. Their support for students and teachers fluidly improvises based on word predictions and not always on validated models of domain knowledge previously described. This leads to risky inconsistency in the information AI tutors provide and the pedagogical approach they pursue. Technology developers have sought to counteract this risk with careful constraints to ensure their systems are reliable.

When they work, these systems hold the potential to support deep learning and motivation, but there remains risk that AI technology produces misinformation or helps students circumvent their learning. With a great distance still to go, intelligent tutoring systems and generative AI have nonetheless incrementally improved their ability to support learning.

The key in all of these efforts has been to try to build tutoring systems that can match the difficulty or mental effort required to do the work with students' capacity. Visually, we might understand the relationship as depicted in figure 4.1. When cognitive load exceeds student capacity, the work is too difficult and can result in confusion and anxiety. When student capacity exceeds cognitive load, the work is too easy. While this may seem less problematic, in reality, boredom is more dangerous to learning than confusion, as it can lead to disengagement. In different fields, this matching of difficulty with ability is known by different names: cognitive load (Sweller, 1988), the zone of proximal development (Vygotsky, 1978), desirable difficulty (Bjork, 1994), productive struggle (Murdoch, English, Hintz, & Tyson, 2020), and flow state (Csikszentmihalyi, 2000).

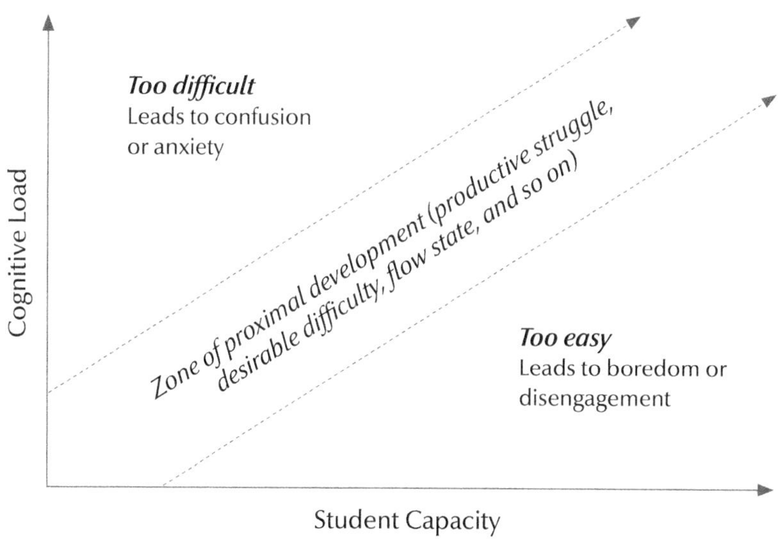

Figure 4.1: *Matching the difficulty of work with student capacity.*

In teaching, we already use technology to scaffold heavy cognitive load, reducing the difficulty of the learning to align with student capacity. Examples include creating concept maps of complex ideas, on paper or on a whiteboard, to help students see how ideas relate or using worked examples to demonstrate procedures, before gradually releasing responsibility. Now, with AI, we can go further in scaffolding cognitive load, like by using AI to individually level difficult texts to each student's reading level to help them access challenging content or by using it to walk students step-by-step through complex problems, prompting them with guiding questions along the way.

When evaluating learning experiences, psychologists also differentiate between *intrinsic* cognitive load, the mental effort necessary to achieve desired learning objectives, and *extraneous* cognitive load, the mental effort necessary to complete a task but tangential to the learning outcomes (Lovell, 2020). In the previous examples, whiteboards, worked examples, and AI tools for adjusting reading levels are all technologies that *scaffold intrinsic cognitive load*, thereby matching the difficulty of the work with the capacity of the students.

Technology can also keep cognitive load matched with student capacity by *minimizing extraneous cognitive load* that might be overwhelming. Examples from simple technologies include using a calculator on a physics problem to reduce computational steps, allowing students to stay focused on conceptual understanding, or using a shared organizational system inside of a learning management system to reduce the organizational structures students need to navigate to obtain course information. Now, with AI, we can go further to help students reduce extraneous cognitive load. Students in humanities classes might use AI to automatically generate citations so they can focus on evaluating and integrating sources instead of formatting citations. Similarly, in a science class, if a project calls for students to make a poster about a topic they are researching, AI can design the poster layout so students can focus their energy on research, not graphic design. In a graphic design course, however, the roles might be flipped: Students could use AI to generate information about a topic, and then the students could design the poster based on that information, thereby applying their cognitive effort to tasks intrinsic to the learning objective and delegating extraneous cognitive load to AI. See table 4.2 for approaches to addressing cognitive load.

Table 4.2: Cognitive Load Theory and Its Implications

Type of Cognitive Load	Definition	Approach
Intrinsic	Effort devoted to processing the complexity of the material itself	Scaffold intrinsic cognitive load that is beyond students' capacity.
Extraneous	Cognitive processing necessary for doing the task but tangential to the learning outcomes	Minimize extraneous cognitive load to enable students to focus on learning objectives.

Distinguishing which tasks are intrinsic and which are extraneous has been shown to improve learning and retention of lesson objectives (Sweller et al., 2019), which can then increase students' self-efficacy and motivation (Locke & Latham, 1990). It's a reminder that being clear about learning objectives allows teachers to more readily recognize what work is necessary and what work is not. This then, in turn, provides a framework that AI can plug into and help target the appropriate amount of challenge for students.

Most research on tutoring systems and work with generative AI have focused on cognitive outcomes: test scores, speed to mastery, or knowledge retention. But learning, as this chapter argues, is more than cognitive. And without attention to motivation, agency, and other characteristics of the heart, even the best-designed systems will continue to fail in the first mile of student engagement. AI, however, is making advances in this new territory, too.

Heart: AI Gets Social and Emotional

Less than two years after the launch of ChatGPT, the second most used web-based generative AI app was not a productivity tool, a homework helper, or an image generator—it was an AI companion app called *Character.AI*. More than seven million times a day, on average, people were already engaging in conversation with digital companions on Character.AI—for fun, counseling, romance, and more (Moore, 2024).

Bots like these have been called *synthetic relationships*, *AI companions*, *social AI*, and more. No longer science fiction sideshows, synthetic relationships had already gone mainstream—and not just with the boys. "The AI boyfriend business is booming," trumpets one headline (Morrone, 2024). "A growing number of women," the article begins, "are seeking connection and comfort in relationships with chatbots—and finding their approximation of empathy more dependable than many human partners' support."

AI companions had arrived, and they have quickly grown pervasive since. Profiles in *The New York Times* and elsewhere highlight retirees, both male and female, building relationships with AI companions for company, intellectual conversation, and love (Dzieza, 2024; Hill, 2025). Kids, as they often do with new technologies, arrived en masse, too. By July 2025, a report by Common Sense Media finds that in a sample survey of over one thousand teens aged thirteen to seventeen, 72 percent have used AI companions (Robb & Mann, 2025). These teens reported primarily using AI companions for entertainment, but they also reported using them to practice difficult conversations they would have in real life. Further, even while they overwhelmingly prioritized human relationships over AI relationships, a third of teen users said they prefer having difficult personal conversations with their AI companion instead of with a human.

For better and for worse, many humans find AI companions emotionally engaging. And it is not just light conversations. One AI-companion company called Replika released an early study of over one thousand college students in which the chatbot reportedly halts suicidal ideation in 3 percent of participants without being directed or taught to do so (Maples, Cerit, Vishwanath, & Pea, 2024). Some people have felt healed by AI companions. Others have become tragically obsessed. In the fall of 2024, for example, one high school student committed suicide following an obsession with his AI companion (Roose, 2024).

To address growing mental health challenges in society, some companies have released counseling tools like Woebot, which are designed to provide emotional resources for individuals when they might not have access to a counselor. New technologies like these and Character.AI signal a new step in the twenty-five-year-old research area known as *affective computing*—computing that can interpret human emotions and responsively engage with them. Advances in AI have marked a new era in which technology meets individual human social needs at greater scale, with both the opportunities and risks that come with it.

If a significant part of the first-mile problem is the gap between the social-emotional readiness of a student and their learning experience, then socially and emotionally engaging AI technology may be able to support unmotivated students, help them manage complex feelings, and foster feelings of confidence. According to students, this is what their best teachers do. In a 2024 survey, Gen Z youth reported that the number-one characteristic of the best middle or high school teacher they ever had was that they "cared about [them] as a person" (Gallup & Walton Family Foundation, 2024, p. 15). Students engage deeply and succeed not only because adults have high expectations for them, but also because they feel adults care for, believe in, and support them (Gallup-Purdue, 2014). Put more starkly, Isabelle C. Hau (2025) writes in *Love to Learn* that "the quality of interactions between teachers and children is a better enabler of academic success than the content of what is taught, the way it is taught, or even the number of children in the classroom" (p. 231). For the vision of AI learning assistants to succeed, students must feel that AI learning assistants understand them like their best teachers do—not simply for their cognitive mastery but for all their emotional complexity.

Are AI teaching assistants that understand and support human emotions a utopic or dystopic vision? To some, the prospect of a robot calming down a human seems a dangerous reversal of the appropriate power hierarchy: Humans should be driving behaviors in the machines, not machines driving behaviors in humans. But if a person in distress can speak with a robot and calm down, or if a person who is frustrated can feel encouraged, or if a student who is unmotivated can be motivated to learn, then this development may mark the arrival of new approaches to machines supporting human learning and relationships. It is a delicate balance with associated risks, and it requires critical attention to how we live and work with technology.

Educator Michelle Culver, founder of The Rithm Project (as in algo*rithm*), separates AI behavior into four quadrants (see figure 4.2; The Rithm Project, 2024). One axis divides AI into a spectrum ranging from prosocial to antisocial based on whether it strengthens or weakens human relationships. The other axis describes AI as on a spectrum from being a tool that you can use to help you perform a task to a companion that provides support or completes tasks independently.

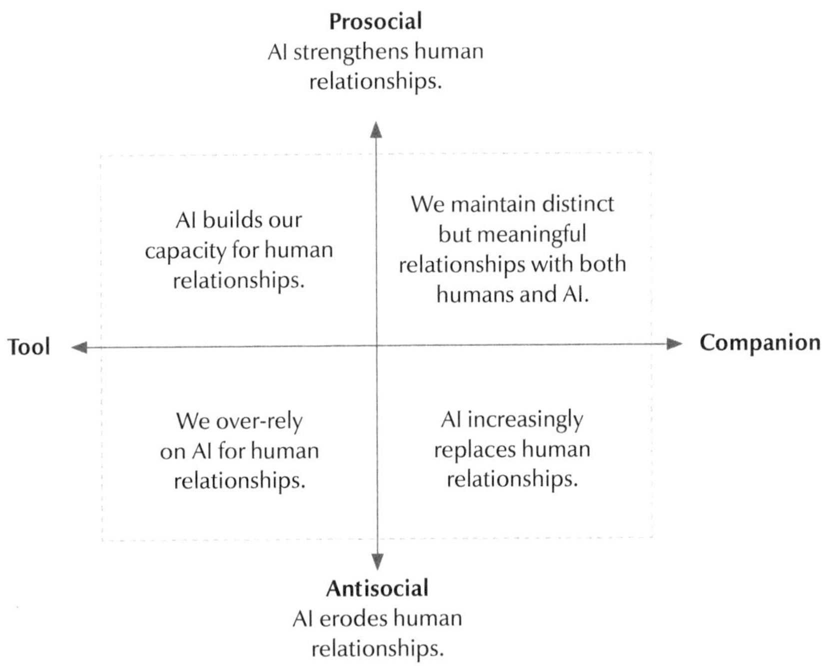

Source: The Rithm Project, 2024.

Figure 4.2: *Adaptation of The Rithm Project's AI quadrants.*

If AI learning assistants are prosocial (staying within the top two quadrants), then they may be better suited to support student mastery without human teachers needing to be present. This doesn't mean the teacher is no longer necessary. On the contrary, the technology strengthens the relationship between students and the teacher. The work that students perform with an AI learning assistant is what sets them up for success when they return to class and engage in meaning making facilitated by the teacher.

The key shift is that, if AI can close more of the first mile through affective engagement, then students will have more opportunities to gain mastery on their own, thereby enabling teachers to increase the time they spend helping students make meaning from what they learn at home. Class time may not suddenly become a seminar every day, but it may incrementally focus more on collaborative and embodied experiences.

Full human experience is broader and deeper than our cognitive processes. When AI engages this broader human experience—starting with our social and emotional needs—it reaches a gate through which it must pass to more successfully support students. To more closely approach the vision of an AI tutor, AI must be designed to treat humans as humans, strengthening our social and emotional richness, not ignoring or denying it. Without this step, each new tool will be just another zombie creation.

Hands: AI and the Body

Around 500 BCE, the poet Simonides of Ceos was at a luxurious dinner party that would change the lives of people for thousands of years (Cicero, 55 BCE/1942). Simonides had written and delivered a poem in honor of his host, a noble named Scopas. Later in the evening, according to Cicero's telling of the story, Simonides was mysteriously called outside. He exited the stone banquet hall, but when he left, he found no one there. At that moment, an earthquake struck, and the roof of the banquet hall collapsed, crushing everyone inside—including Scopas—but sparing Simonides, who fortuitously was outside. In the aftermath of the earthquake, family members did not know who was who under the rubble of the building. Simonides discovered that by recalling the shape of the dinner table, he could remember everyone who was in attendance based on where they had been sitting, and family members could therefore find the remains of their loved ones. Of all the events that transpired that evening, it was the method of remembering who was there that most changed the future.

Simonides's method of imagining the physical layout of the banquet hall has been adapted into what is now called the *method of loci*. By placing thoughts and ideas into imagined physical spaces, the same way Simonides imagined the dinner table, people can remember nearly limitless information. For hundreds of years, orators would use this technique to memorize lengthy speeches. Instead of remembering seats around a table, as Simonides had, orators would imagine vast palaces in their minds, and they would place sections of their speeches in different rooms of these vividly imagined buildings. Then, when delivering a speech, they would mentally walk through the palace, strolling from room to room to trigger the memory of each section of their speech.

What the story of Simonides taught us is that the brain exists and has evolved in three-dimensional space. As a result, thinking and memory are deeply connected to movement and the body. The physical, sensory, and spatial world is the infrastructure of our thought. Our embodied experiences in the world are why we look "back" at something, "weigh" options, "grasp" ideas, or think "through" topics. Our thinking exists in a world of metaphor (Lakoff & Johnson, 1980), like the first-mile, last-mile problem referenced earlier. We use the physical world to give "shape" to our thinking. This helps us realize that a brain in a jar—without senses or a body, as often depicted in science fiction—would have no frame of reference for its thoughts. Our language depends on embodied experiences in the world. We express love in the shape of a heart—a bodily organ. We represent time, which we cannot see, as a line on a page. Even the word *topic*, an abstract term, conveys physicality, as it comes from the Greek word *topos*, meaning *place*.

This embodied understanding of how the brain works extends to how we express ourselves. Our words transmit meaning, but our bodies often transmit much more. In two studies, psychologist Albert Mehrabian found that when body language or tone of voice contradicted spoken words, people found greater meaning in the body language or tone of voice (Mehrabian & Ferris, 1967; Mehrabian & Weiner, 1967; Sato, Shimokawa, & Minato, 2025). We see this every day. Who hasn't been skeptical when a child (or partner), when asked how they're doing, flatly says they are "fine" but avoids eye contact, frowns, and skulks away?

Sense as a kind of thinking is inseparable from sense as a kind of feeling. We learn when we feel. We think when we move. And our learning happens in our bodies as much as in our brains. It shows on our faces. It sounds in our voices. It is held in our posture.

Science writer Annie Murphy Paul was working on a book about the cognitive science of learning when her research led her to a revelation:

> Our culture insists that the brain is the sole locus of thinking, a cordoned-off space where cognition happens. . . . Many tomes have been written on human cognition, . . . but they are limited by their assumption that thinking happens only inside the brain. . . . Yet a parallel narrative has existed in front of us all along. . . . We think best when we think with our bodies, our spaces, and our relationships. (Paul, 2021, pp. 8, 9, 10)

After this revelation, she dramatically shifted the focus of her book from the cognitive science of learning to how our bodies, surroundings, and relationships affect our thought. Its title? *The Extended Mind.*

Now, in line with Paul's findings, AI is not only working with us cognitively but also engaging with us physically in two ways: (1) by reading our physical experience in the world and (2) by itself taking on an embodied experience through robots.

First, as a tool that reads our embodied human experience, AI shows a growing capacity to recognize emotions in facial features and body posture. Whether and how this can be put to effective and responsible pedagogical use is yet to be seen, but the technology is increasingly capable. In early studies, AI systems show promise in detecting affective states like confusion, boredom, or frustration through nonverbal cues like slouched posture, head tilt, gaze aversion, and microexpressions (D'Mello & Graesser, 2012; Happy, Dasgupta, Patnaik, & Routray, 2016). This suggests that machines might be able to perceive and respond to the physical signs of learning, just as skilled teachers might. An early trial by Naixin Zhang and Wai Yie Leong (2025), for example, reports increases in student engagement when AI tools detect emotions and trigger timely, adaptive feedback. But this research is young, and there are significant ethical questions about privacy and bias. Should these be resolved, then our learning technologies will have greater ability to attend and attune to the embodied nature of our minds.

Second, as an embodied tool for learning, AI can increasingly engage with our own spatial and kinesthetic selves. In early studies, for example, robots using their physical presence, gaze, and gestures foster increased attention and engagement, especially among young learners (Belpaeme, Kennedy, Ramachandran, Scassellati, & Tanaka, 2018). Similarly, robots have elicited kinesthetic movement in students that helps them learn abstract concepts through the robot's modeled movement (Di Canio, 2021).

While this kind of research is young, it is possible to imagine what this might look like in use: A student is studying geometry at home and is using an AI tutor robot to help them understand the mathematical concepts they are learning. The AI tutor robot demonstrates angles with its arms and invites the student to mirror the movement. If it detects signs of confusion in the student's facial expression or tone of voice, it offers a new approach. When the student succeeds, the robot raises a hand for a high five, which the student gives. The interaction is multisensory, emotional, and physically experienced.

We are far—both technologically and culturally—from this kind of physicalized experience being widely effective and equitably distributed, but ongoing research does shed light on these important possibilities for and elements of human learning.

We learn not just with our heads but also with our hearts and our hands. So, as AI technologies converse more and more fluently with us in cognitively enriching ways, interact with us more readily on a social and emotional level, and increasingly read our expressions and engage with us kinesthetically, their ability to advance student learning outside the classroom continues to expand, too, because they have engaged with us in ways that treat us like humans, not like machines.

And if they succeed, they will likely have achieved the long-standing vision of an AI tutor. Which then leads us to ask, "Will we still need teachers, even then?"

The Ends of Education

It's easy to forget that learning isn't the purpose of education.

The purpose of education is to enable people to participate in society, pursue meaningful work, and live fulfilling lives, both individually and collectively. Learning is the means. It is the path to a good life, however that might be defined.

When we think that all our learning can happen with an AI tutor, we forget that our lives are lived among other people. What we learn matters because of how we bring it to the people around us: our loved ones, friends, colleagues, or fellow inhabitants of the world. What we learn may have personal meaning to us when we learn it, but it—and we—become part of the fabric of society when we bring our learning and ourselves to others.

Our different teaching philosophies express this social imperative in different ways, but they share it nonetheless.

- For **pragmatists**, education develops the collaborative skills needed to thrive in the workplace. They recognize that organizations are collaborative networks that also serve a broader role in society.
- For **academics**, education advances scholarship built on generations of work by other people. They value navigating diverse academic viewpoints and recognize that the human institutions that make scholarship possible are themselves complex social systems.

- For **humanists**, education deepens our understanding of ourselves and others. They see human nature itself as social and learning about others as central to the richness of the human experience.
- For **socially oriented educators**, education shapes society toward greater equity and justice, which are inherently oriented toward relationships in communities. Their objectives can only be accomplished through dialogue and shared action.

Teachers are irreplaceable because students will always need to make meaning from what they learn. And making meaning in school means convening with other students under the guidance of a teacher. Together, we check and reconcile our different individual interpretations, develop visceral understandings of how what we learn affects the lives of others, correct our own misconceptions, update our beliefs, and shape how we choose to act in the future because of what we learn from other people. The value of learning is not in the abstract; instead, it is in how all that we learn leads us to conversation and action with others. This will always be the case, so long as we live in a society of other people.

Society demands an education that is greater than our individual learning alone. It will be decades—or longer—before AI tutors can engage us in a generalized way across head, heart, and hands. But even as we look ahead to that future, we must recognize that learning alone with a machine is an incomplete education.

Social Understanding and Intellectual Humility

As authors of this book on AI, we are both frequent users of AI tools, most significantly for research and exploration. In our work with AI, we have found it to be an enormously compliant assistant. It is available at any hour, it responds with consistent enthusiasm and positivity, and when we challenge its findings or offerings, it dutifully revisits what it has provided and makes any adjustments we ask of it. It holds no perspective strongly and is ready to embrace the direction we suggest—every time. Both of us have learned much from our work with AI, and our collaboration with AI has been remarkably informative, enriching, and easy.

And then we return to our meetings with each other.

As teachers, researchers, and writers, we each have strong opinions about the directions our work might take—and they are not always the same. We are very different from AI. Rather than dutifully complying with a suggested change, we challenge each other's ideas. We reference not only evidence to support a claim but also our experiences. Other times, we might tenaciously hold on to an idea, even if we have no evidence, because we have a hunch that it is right but can't quite articulate it yet—or maybe ever. Sometimes, the hunch is right, and other times, it isn't, but we hold fast until we know or are convinced otherwise. Working with each other is much more complicated than working with AI.

But, mercifully, we both deeply recognize the value that comes from collaboration. We understand that our divergent perspectives and intuitions on topics are fruitful. We recognize that from productive struggle—through these differences—comes new insight, and from creative tension comes new ideas. We know that The Beatles wrote better music together than any of them did alone. For this book, we believe the meaning that we have found at the intersection of AI and education—and the way we organize and articulate it—is far richer and more textured because of the discussions that we have had together. It has pushed both of us, at different times, to revise some of our own beliefs, adopt each other's, or even codiscover and codevelop new beliefs and frames of thinking as a result of the conflict and convergence of our different perspectives.

This is the essence of meaning making that is irreplaceable in a time of AI. It requires bringing what we learn to other people, as well as humility, persistence, curiosity, interpersonal understanding, and more. These are dispositions that can always be improved and require ongoing practice and guidance.

When students learn from AI tutors that engage with them through the head, heart, and hands, they will be able to access more information and develop more skills than they could before, but they will still need to convene, with expert teachers guiding them in making sense of what they learn and in building the bonds with their fellow classmates that are necessary for a healthy society.

Pedagogy in a Time of AI

The case that we have made so far in this chapter is that AI tutors will be most effective when they can engage students through the head, heart, and hands. But this will take time. The flipped classroom did not suddenly move the full balance of knowledge and skill acquisition out of the classroom, but it did mark a shift. Now, with AI, the balance is shifting further.

As this fuller version of independent student learning grows more successful and as students more readily master content and skills on their own with an AI learning assistant, teachers can plan their classes with less time reviewing for mastery and more time facilitating collaborative and embodied experiences for making meaning. What this might look like is the focus of the remainder of the chapter.

Lesson Design Leveraging AI Learning Assistants

What might deeper engagement look like when an AI learning assistant can support students to work more independently? In table 4.3, the left column shows examples of how homework might look different with an AI learning assistant, and the right column shows how class might be shaped if students have achieved greater mastery on their own.

Many of the AI capacities described in table 4.3 are available today. Others will take time to develop. Notably, none of the examples describe how AI assistants might respond to unmotivated students, distracted students, or students who otherwise encounter unanticipated obstacles to their learning. Addressing those situations requires the social-emotional or embodied domains of learning noted earlier in the chapter—and these are, for now, the domain of teachers. In lesson design, while we can already increase the balance of what students are able to do on their own, the teacher remains the best assessor of

Table 4.3: Shifts in Student Work and Class Time With an AI Learning Assistant

Homework and an AI Learning Assistant	Class Time
History: A student reads primary and secondary sources for an event in the civil rights movement and debriefs them in conversation with an AI learning assistant. AI answers the student's questions, poses questions back to the student, and explores the people involved or the contextual history with the student.	The class discusses tensions at play in the historical event and how they relate to tensions at play in the present that students may be affected by.
Biology: A student reads a short case study on CRISPR and gene editing and uses the AI learning assistant to map arguments for and against its use. AI offers different ethical perspectives the student may not have considered and engages the student in conversation about the values related to each.	Students and the teacher have a structured discussion on the ethics of gene-editing technology and what its impact might be in their own community.
Physics: A student plays with simulations of objects colliding. An AI learning assistant engages in conversation with the student, inviting their predictions of what will happen and debriefing how the predictions matched the behaviors.	Students and the teacher perform controlled collisions in class with real-world objects to demonstrate physics principles in a more multisensory way. Then, they engage in discussion about how these principles apply in sports, dance, or road safety, like the stopping time for cars and trucks of different sizes.
English language arts: A student reads a chapter in *The Outsiders* at home and engages with an AI learning assistant by asking questions about the characters involved. Then, the student and AI role-play different characters and scenarios in the chapter.	Students and the teacher discuss together their experiences of inhabiting the perspectives of different characters and enact them with each other, seeking to develop empathy for the ways people experience a moment differently.
Mathematics: A student walks through a three-act mathematics problem about the cost of a shower versus the cost of a bath (Meyer, 2011), and then, with guidance from the AI learning assistant, gathers and prepares information about their own shower or bath. AI answers the student's questions as the student proceeds, assessing their progress.	Students and the teacher together compare results across the class, calculate average costs and rates, and discuss and calculate the implications of the cost and water supply across larger communities.

continued →

World language: A student practices a new conjugation in a foreign language, and an AI learning assistant creates low stakes practice opportunities for retrieval practice, in writing and in conversation with AI.	Students interview each other and practice using the new conjugation on the fly, in new contexts they didn't anticipate.
Physical education: A student sets goals and logs their daily physical activity (such as push-ups, distance running, or flexibility exercises). An AI learning assistant offers encouragement, provides feedback on their progress, and suggests adjustments based on video footage.	Students compare progress in small groups, discussing patterns and areas for improvement. The teacher leads a session on interpreting fitness data, helping students adjust their personal goals and strategies.

when students need freedom to thrive and when they need connection to grow.

Self-determination theory (Deci & Ryan, 1985) tells us that the greater the work's relatedness, meaning the more it connects students to others, the greater their intrinsic motivation will be. As class discussions increasingly involve collaborative meaning making, low motivation may be less and less of a challenge. For students, the work they do with AI outside of class may make the experiences they have with teachers and peers in class even more irreplaceable.

Lesson Design for Leveraging Cognitive Load

Not only can designing lessons for students to use AI shift the balance of where students are gaining mastery, but it can also help scaffold cognitive load in student learning. The result is not that students exert less cognitive effort but that with the same cognitive effort, they can reach new heights. AI scaffolds students' ability to do more.

What might this look like in your classroom? Table 4.4 explores a variety of ways in which teachers might leverage AI to scaffold more ambitious student learning. The left column describes scaffolding strategies many teachers already use, and the right column shows an example of that strategy in use, both in its current form and with AI, to support new learning.

Table 4.4: Strategies to Scaffold Intrinsic Cognitive Load

Scaffolding Strategy	AI Scaffolding Example
Segmenting content	**Current:** Teach mathematics in step-by-step procedures. **With AI:** Have AI-powered apps deliver bite-sized microlearning modules.
Pretraining and activating prior knowledge	**Current:** Review key terms before an advanced biology lesson. **With AI:** Use personalized AI-generated preassessments to identify and fill knowledge gaps.
Chunking information	**Current:** Provide historical facts organized by economic, political, and cultural themes. **With AI:** Have AI suggest which fact to consider next and offer some relevant context for it.
Using worked examples	**Current:** Model solving a chemistry equation through a detailed demonstration. **With AI:** Use AI to generate worked examples that are tailored to a student's individual misconceptions or errors.
Using analogies and examples	**Current:** Explain electricity flow by comparing it to water in pipes. **With AI:** Have an AI-powered analogy-generation system provide personalized analogies based on learners' interests or backgrounds.
Incorporating visual and auditory dual coding	**Current:** Combine diagrams of biological processes with audio explanations. **With AI:** Have AI create multimedia explanations that synchronize diagrams, animations, and narrated explanations.
Sequencing and pacing	**Current:** Ensure students master foundational mathematics before advancing. **With AI:** Use an AI-based adaptive-learning platform that automatically adjusts lesson pacing based on learners' progress.
Focusing attention	**Current:** Use bolded key terms or clear summaries to signal important content. **With AI:** Use AI-driven content summarization or highlighting tools to spotlight essential ideas and minimize distraction.
Interleaving and task variation	**Current:** Alternate types of mathematics problems to reinforce understanding. **With AI:** Have AI provide varied problem sets to reinforce and interleave skills.

These approaches to using AI empower students to do more in service of pursuing the learning objectives we have for them. If students are able to engage more deeply in this way outside of class, it further invites us to revisit how we use our class time with deeper engagement in mind.

The Push and Pull of Good Pedagogy

Good teachers and tutors are always reading their students. They know when to give students space and when to reach out to them. It's a subtle improvisational skill, and it is also an essential one.

Technology is not good at this yet. Our devices either barrage us with push notifications or wait passively for us to pull utility from them. The great lesson from research on learners' cognitive control while multitasking with media (Ophir, Nass, & Wagner, 2009) is that the *push* notifications adolescents receive from social media apps on their mobile devices fragment their attention. The great lesson from the rise and fall of MOOCs (massively open online courses) is that the overwhelming majority of people do not persist in *pulling* learning from technology (Holt, 2024; Reich & Ruipérez-Valiente, 2019). One of the central challenges, therefore, for the development of AI learning assistants will be calibrating the right balance between AI proactively prompting students and more passively following where the students take it. It will mean knowing when and how to reach out to students, then when and how to follow them.

What this chapter endeavors to show is something that great teachers already know: One of the greatest sources of motivation for students is the relationships in their lives. They care about work when they feel their teachers care about them. They care about learning when it helps them better relate to their peers. Our pushing and pulling as teachers—our delicate pedagogical dances and designs—serve to keep students engaged in their relationships with each other, with the curriculum, and with us.

For many teachers, this dance is entirely intuitive. It is built on relational intelligence, on the ability to read students' engagement, and more. Like a dance, it is an art. If the science of learning reveals to us the levers of the head, heart, and hands, the art of teaching is knowing which levers to pull and when.

Review and Next Steps

Great teachers do more than design effective learning experiences—they build relationships with students, recognize and respond to how students feel and present themselves in class each day, and strengthen students' relationships with the curriculum by leveraging their own mastery of the material and the available tools to affirm what students already know and to deepen it further.

Technology can't do all this the way teachers do. But, incrementally, it can do more and more of what teachers do, freeing teachers up to do more still (and maybe also sleep and eat). As technology increasingly becomes a part of our social and embodied human experience, we will have more opportunities to leverage it to more effectively support student learning.

This blurring of boundaries between human and machine is accelerating, and it carries substantial risk. In a time like the present, frameworks like The Rithm Project's prosocial and antisocial spectrum ought to live at the center of our thinking and work. This means going beyond the call for the simple presence of a human in the loop and further asking, "Do new technologies actually *strengthen* human relationships? Are they *prosocial*? How can AI learning assistants not replace—or even simply supplement—relationships with teachers but actively deepen them instead?"

Still, while an AI learning assistant that helps us understand is a great benefit, understanding alone is shallow learning. Putting understanding in action, driving learning to the higher parts of Bloom's taxonomy and enabling students to demonstrate their learning, is the real work of what we do in school. An AI *learning* assistant helps students access understanding; an AI *doing* assistant helps students demonstrate their understanding. Students *learn* by *doing*, after all. What students *do* with AI is the focus of chapter 5. But first, here are some next steps to consider for educators, school leaders, and those working in the technology industry.

For Educators

The following are some considerations for educators related to AI for student learning.

To understand the student experience, try using AI to master a new content area, concept, or skill. Revisit the experience of learning something challenging and new by engaging with an AI tool to discover and seek to master a new topic. Choose something in your subject area that you feel less confident in or want to explore more. What works when you engage with AI to learn something? What do you find compelling? What do you find frustrating? What do you wish it could do that it can't? Explore and then reflect: How might your students experience this same opportunity? What might work for them? What might not?

Consider which areas of your curriculum students might learn on their own and what they might achieve working with peers or a teacher. Part of school is learning a critical mass of content. Other parts require practicing skills, some of which students can practice at home and some of which they may require a classroom or peers for. Still other parts of school are about social meaning making. How might the work in the courses that you teach separate around these categories? What do students need a classroom or peers for, and what do they *not* need a classroom or peers for?

Think outside the box of current constraints and determine how class time would change if students could master more content and skills at home. As you are able to move more content acquisition and basic-skill acquisition outside of class time, how might you shift the balance of what you do in class? More labs? More discussions? More projects that put knowledge to use? How might your course objectives be fulfilled if you allotted more time for students to work together?

For School Leaders

The following are some considerations for school leaders related to AI for student learning.

Host conversations about student social-emotional engagement with AI. The study of social use of AI is in its infancy, but it must mature quickly. What do adults know about student use of AI companions? What do faculty know about AI companions and synthetic relationships? As social-emotional engagement with AI spreads to schools—perhaps first intentionally through supplemental therapy bots and then more surreptitiously through the embedding of social-emotional behaviors in academic AI tools—how can teachers and school leaders enter this new era with eyes wide open?

Devise guidelines or policies for social engagement with AI tools. As conversations about AI companions and synthetic relationships unfold, what boundaries might emerge that could serve as guidelines or policies for the school? What do students need to know about AI companions for healthy social development? What is a safe way to introduce the topic without inadvertently causing more harm?

For Technologists

The following are some considerations for those working in the technology industry related to AI for student learning.

Move slowly when exploring social AI. Acceleration of frontier models is inevitable, but acceleration of social AI for students is not. When we "move fast and break things" in the field of social AI, the "things" we break may be higher stakes than in other fields of technology. While social-emotional engagement will be a powerful next step for educational technology, it is also fraught with risks and is a domain in which proceeding with caution is necessary.

Work with a variety of skilled teachers when developing pedagogical modes for AI. Training AI to work like a teacher means much more than giving hints instead of answers. Most meaningful teaching isn't about reaching answers, anyway—it's about students' mastery of skills and textured understanding of challenging questions. Expert teachers engage a wide variety of pedagogical approaches. They think creatively and responsively about every class and every student. Good pedagogy doesn't mean using visual aids all the time; it means matching pedagogy to content and context. Leaders know that the best leadership style is the one that is needed at a particular moment. Teachers know that the best pedagogical style is the one that is needed for their particular students at that particular moment. This requires science-supported general pedagogy, discipline-specific Pedagogical Content Knowledge (Shulman, 1986), technology-enabled Technological Pedagogical Content Knowledge (Mishra & Koehler, 2006), and other approaches to engaging students in teaching and learning. There is simplicity to developing this, but it is on the other side of complexity. It will be through testing a variety of approaches that a learning assistant will be able to effectively support a student.

CHAPTER 5

The Doing Assistant

Empowering Learners to Strategically Manage AI

I hear and I forget, I see and I remember,
I do and I understand.

—XUNZI, CONFUCIAN PHILOSOPHER

Why do we hike? If it's just to get to the summit, it would be much easier to drive up a road. In New England, for example, it is common to see "This car climbed Mount Washington" bumper stickers from those who chose that route. So why do tens of thousands of other people choose to make the arduous trek by foot, anyway, in all kinds of weather?

If it's not the view, perhaps it's the exercise. Sure, we could exercise by running on a treadmill, but hiking is definitely a more interesting way to burn calories and strengthen muscles. Still, not all hikers are working up a sweat; some walk quite casually up the mountain, stopping and resting whenever they feel like it. Maybe, for those people, it's about the little discoveries they make along the way: the animals, brooks, and vistas on different promontories. Or maybe, it's about the companionship of the friends they are hiking with.

Now let's imagine a new technology was invented: teleportation. If Mount Washington installed a teleportation device that allowed people to be transported instantly from the bottom of the mountain to the top, how many people who drove would take advantage of it? How many of the hikers? No matter how easy it is to get to the top, some people will always choose the hike—not because it's easy but because it's hard, as John F. Kennedy famously said.

For those of us who love hiking, the process—the hike—*is* the goal. The journey *is* the destination. The summit only marks that we accomplished what we set out to do. We hike because we value what happens along the way. We have grown stronger. We have developed navigational skills. We have deepened our knowledge of the mountain. We have built relationships with the people in our group and others we have met. We feel more alive on the hike. The view is the bonus, not the purpose.

In this metaphor, hiking is of course the productive struggle of learning, and the teleportation device is AI. Will our students choose to teleport, or will they hike? This chapter explores that choice and what it means for students to produce meaningful work outside the classroom in a time of academic teleportation. But it's not just about insisting that every student hike; it's also about when, why, and how they might choose to explore mountains when teleportation devices can take them anywhere.

But before we explore working with AI, we believe there's also something deeper. Underneath the fear some teachers have about how students are using AI is a more subversive problem that affects how students approach AI to begin with.

The Agency Crisis

When a teacher asks a question, students try to give the right answer. When a teacher provides detailed instructions and a rubric, students try to carry them out and are graded on their adherence. What starts as an important effort to support students and make sure expectations are clear turns into a pedagogy that overprotects students from failure (Holt, 1964). If the medium is the message, then the messages conveyed in these kinds of experiences include (a) knowledge and authority sit with the teacher, (b) success means obeying detailed instructions to create prescribed products, and (c) external affirmation matters more than genuine understanding or curiosity.

This diminishment of student agency is apparent in other patterns in our society that also strip young people of agency.

- Highly involved parenting reduces children's decision making.
- Algorithmic feeds help us choose what to watch, read, and buy.
- Optimization culture provides "best practices" for everything.
- Standardized testing defines singular success paths.
- Social media outsources self-worth to external metrics.

In short, students' environments today might be training them to ask, "Where do you want me to go?" instead of "Where could I take this?"

Maya experienced this directly her first year teaching seventh grade. She had assigned a project that gave students the latitude to set their own goals, but many were confused by what she meant. She explained that they could choose the direction of their work, adding, "If you don't know where you're going, how will you know when you get there?" The question was rhetorical, but one student answered anyway: "When my mom pulls the car over and opens the door." No one giggled. Students nodded. This was not a joke. Students have become excellent at executing others' visions but struggle to develop their own or even understand what it would mean to have one. And it's been happening for decades.

When sociologist Jean Anyon began observing a set of New Jersey fifth-grade classrooms in the late 1970s, she expected to document variations in lesson plans. Instead, she uncovered variations in messaging. In her landmark article, "Social Class and the Hidden Curriculum of Work" (Anyon, 1980), she showed that schools were not merely teaching arithmetic and grammar; they were subtly teaching students their social roles. She observed schools at four different levels of socioeconomic status and documented the messages each school conveyed to students. In schools serving the "working class" (low socioeconomic status), students were rewarded for following rules exactly (and punished for not doing so), setting them up to be obedient workers. In schools serving the "middle class," students were taught to do assignments neatly and completely, training them to be reliable technicians. In "affluent professional" schools serving the upper-middle class, students were encouraged to figure out why something works and present their original ideas. This set them up to be creative professionals who are able to improve on existing systems. She coined the schools serving students at the highest level of socioeconomic status as the "executive elite." In these schools, students were taught to critique the systems they were in, redesign them, and defend their choices. This prepared them to be high-level decision makers.

Other scholars have expanded on Anyon's work, similarly showing how some schools in low-socioeconomic-status environments focus on compliance with strict behavioral scripts (Golann, 2015), while schools in middle-class and prestigious settings often enable self-advocacy and leadership instead (Calarco, 2018; Cookson & Persell, 1985). But even the elite, it turns out, may not be immune. Former Yale professor William Deresiewicz (2014) argues in his book *Excellent Sheep* that elite education trains students to be "great at what they're doing but with no idea why they're doing it" (p. 3). So, the agency crisis in education affects us all, and, as with so many problems in education, it affects the most vulnerable the most.

With the challenges we face as a society, this is unacceptable. We need students to grow into citizens who develop and carry out moral ambitions greater than themselves (Bregman, 2025). An eighteen-year-old who learns to hold conflicting evidence without flinching, use AI and critique its answers, and persist through multiple iterations becomes the adult who can thrive in a more complex world—coordinating wildfire responses, redesigning voting systems for greater participation, or launching the social enterprise we didn't know we needed.

But without agency—without feelings of ownership over their futures—students are more likely to take shortcuts instead (Kanat-Maymon, Benjamin, Stavsky, Shoshani, & Roth, 2015). This means that when every student has an assistant, if they don't participate in a culture of learning or believe the *process* of their schoolwork is the *purpose* of their schoolwork, then they may be more likely to use AI just to get an answer—just to make a *product.* It is their feelings of agency that will—at least in part—determine whether they choose to teleport to the top of the mountain or hike.

In 1960, writer John Steinbeck drove around the United States in a customized pickup truck, visiting small towns and meeting local communities. He wrote up his experience in a 1962 travelogue titled *Travels With Charley: In Search of America*. He traveled on small roads, from town to town, meeting people of all backgrounds and seeing sights of all sizes. But before the trip was finished, he grew weary and decided to take the newly constructed interstate highway system home. It was faster and easier, and it certainly accelerated his return.

But he was also unsettled, and he reflected on what was gained and lost by these new roads: "When we get these thruways across the whole country, as we will and must, it will be possible to drive from New York to California without seeing a single thing" (Steinbeck, 1962, p. 66).

AI, like the thruway, is here. If we are not careful, it is possible for students to complete assignments and not learn a single thing. At the same time, if we are to prepare students for the future, we must prepare them to use AI.

But there is good news—really good news, actually. As is so often the case, the essence of the cure can be found within the poison itself. While AI can exacerbate the agency crisis, it provides entirely new opportunities for student agency at the same time.

How can this be? Of course, there are no true shortcuts in learning. As Daniel T. Willingham (2009) puts it, "Memory is the residue of thought" (p. 58). Every thought we have encodes new neural pathways. Whatever students think about is what they are learning, and if they shortcut thinking, they do not learn. But AI enables us to practice new and different kinds of thinking, opening new and different opportunities.

So, what are these opportunities? What will they *learn* if they use AI? Well, it depends on *how* they use AI.

This means reimagining what we expect students to do when they are outside of class. It also means rethinking how we, as teachers, structure learning experiences in general. So, the two main sections of this chapter are built around two main questions: First, what do *students* need to learn and lead in a world with AI? This will help us determine the nature of our new learning objectives, and it sets us up for the second question: How do we, as *teachers*, design learning experiences so students are empowered to use AI as a tool for deeper learning rather than a shortcut around it?

Students' Needs

The AI at every student's fingertips is an on-demand creative team: a ghostwriter, tutor, illustrator, coder, fact-checker, and more. The student is their leader—their executive.

What's scarce now is not *production* but *direction*. This shift reprioritizes the human capacities we emphasize in schools, promoting key skills that matter for team leadership, like framing better questions, judging what is trustworthy, integrating diverse viewpoints, deciding what is worth building, and persuading real audiences to care. The key insight here is not that AI can complete many current assignments; it is that when AI can produce enormous amounts of work, direction setting and decision making rise as *irreplaceable* human skills.

In this section, we look at what students need to lead in a world of AI assistants. We focus specifically on (1) executive skills for reclaiming agency, (2) AI literacy for ensuring responsible use of AI tools, (3) appropriate scaffolding of skills across grade levels, and (4) essential habits of mind for success with AI.

Executive Skills

When productivity is cheap, direction is priceless. Like teachers in chapter three, students are now executives, leading their own creative teams. And like a CEO surrounded by specialists, the human executive does not need to outcalculate the AI accountant or outrecall the AI researcher. Instead, they do the following.

- **Set a vision:** They frame the problem, name the stakes, and draw the moral boundaries within which the work must proceed.
- **Define and delegate work:** They charge the team to draft, simulate, or forecast—always with a clear brief and a plan for how the output will be judged.

- **Evaluate outcomes and provide feedback:** They look for what is missing, challenge thinking, and ask for specific improvements.
- **Integrate multiple perspectives:** They weave the team's findings with other stakeholder voices and the contextual nuance that no individual fully captures.
- **Determine a course of action:** They weigh trade-offs, take ownership, accept responsibility for consequences, and communicate the rationale to those affected.
- **Adjust and adapt:** They treat each creative production cycle as a feedback loop, refining both the questions they pose and the criteria by which they judge the answers.

In short, while AI provides the executive with data and can draft every conceivable slide deck, only the human can decide which slide becomes the strategy and convince their team why that strategy is best. The student-as-executive turns AI's unlimited productivity into meaningful and actionable next steps.

The metaphor of the executive suggests not only the professional role of a leader but also what psychologists call *executive function* (Center on the Developing Child, n.d.; Diamond, 2013; Zelazo, Blair, & Willoughby, 2016). Executive skills are not the same as executive-function skills, but the work of executives is not irrelevant to the similarly named set of neurological processes.

- High-stakes environments require **emotional self-regulation**.
- Overburdened calendars require **time management**.
- Information abundance requires **filtering and prioritizing**.
- Rapid change requires **cognitive flexibility**.
- Social media requires **impulse control**.
- Systems of workflow require **organization**.
- Moral complexity requires **metacognition**.

When every student is the leader of their own AI team, executive skills become increasingly important. And when they have a greater ability to realize and develop their vision, they have greater agency than ever before.

This is not to say that every learning experience should be AI infused, but it is to say that in an age of AI, students must nonetheless learn how, when, and why to use their cognitive teleportation devices and when to hike.

AI Literacy

An executive has to know their team: its strengths, weaknesses, and how team members work together. Without knowing this, an executive risks assigning work that the team might not be well suited for and end up spending more time fixing problems than making progress. But if an executive knows their team, they can match the right person to the right job, anticipate where their judgment and input will be most helpful, and can generally spend their time (and ensure their team member's time is spent) more efficiently and with better results.

Students must know their AI tools like executives know their teams. They must know AI's strengths and weaknesses and how to work well with it. They must have AI literacy.

There isn't a single agreed-on definition of AI literacy—just as there isn't a single agreed-on definition of media literacy—but there is a core set of overlapping competencies related to AI that consistently appear across academic and policy frameworks (Lee, Mills, et al., 2024; Miao & Cukurova, 2024; Stanford Teaching Commons, n.d.). These competencies are in how AI works, how to use AI effectively, and how to use AI responsibly. We outline each of them briefly here.

- **How AI Works**
    - *Core concepts and terminology*—Students should understand what generative AI is, the types of tools available (such as text, images, audio, and video), and key terms, such as *prompt*, *training data*, and

hallucination. This shared vocabulary makes it easier to discuss AI clearly and compare tools effectively.

- *Training and output generation*—AI learns through cycles of input, feedback, and adjustment, processing datasets to predict what comes next in a sequence, whether that's a word, pixel, or note. Students should grasp that these outputs are not thoughts or facts retrieved from an authoritative source but rather statistical predictions shaped by the data and the training process. This understanding helps students interpret results with an informed eye.
- *Strengths, weaknesses, and current trends*—AI is powerful at summarizing information, brainstorming ideas, and producing fluent text, but it can also produce inaccurate, biased, or irrelevant results. Students should know both where AI excels and where it falls short and have some awareness of how capabilities are evolving. This perspective prevents blind trust in the technology and encourages adaptive use over time.

- **How to Use AI Effectively**
    - *Specific and iterative engagement*—Getting strong results from AI often requires creative prompts with detailed information and relevant context. Students should learn how to give instructions that lead AI toward the kind of output they need, experimenting and refining until the results meet their purpose.
    - *Selection of the right tool for the task*—Not every AI tool is suited for every job. Students should be able to match the capabilities of a given system, whether for drafting text, generating images, or analyzing data, to the requirements of an assignment or project.
    - *Review or refinement of AI outputs*—Even well-prompted AI responses require human review. Students should be able to check outputs for accuracy and appropriateness, then revise AI's work or their prompts to improve the result.

  - *Collaboration with AI*—Using AI well is a kind of collaboration. Students should be able to combine AI assistance with their own knowledge and skills and reflect on how AI influences their learning, decision making, and creative processes.
- **How to Use AI Responsibly**
  - *Bias and fairness*—Students should be alert to the ways AI outputs can reflect and amplify biases present in the training data. Recognizing these patterns allows them to question outputs rather than passively accept them.
  - *Privacy and security*—AI often collects and stores input data, and students need to understand the implications for personal, institutional, and third-party information. They should know what not to share and how to protect their own and others' privacy.
  - *Transparency and responsibility*—In academic work, students must abide by expectations for AI disclosure and take ownership of the final product of their work. AI can assist when school policies permit, but the student remains accountable for the integrity and quality of the work.
  - *Broader impacts of AI on people and communities*—Students should think beyond the individual task to the societal, environmental, and economic effects of AI. This awareness helps them see AI not just as a tool but as a force with real-world consequences.
  - *Misinformation*—Because AI can produce convincing but false information, students must develop habits of verifying AI-generated content with credible sources. This protects the quality of their own work and builds a broader skill of digital discernment.

A foundational mastery of these topics ensures that students know how AI works, how to use it effectively, and how to use it responsibly.

But this isn't something they can learn all at once. Instead, they can learn what is appropriate at each level of developmental maturity.

Appropriate Scaffolding of Skills

An executive typically rises from within an organization, incrementally learning the skills of their trade. Whether in a school, on a sports team, at a research lab, or in a corporate office, an executive accumulates understanding over time in a way that compounds their knowledge.

Children follow a similar trajectory, starting with foundational knowledge and skills that match their cognitive and emotional readiness in lower years and building in sophistication as they grow. For example, most students don't learn to write with a keyboard or begin mathematics using a calculator. They typically learn to write by hand before they begin typing, and they practice arithmetic with blocks and long division long before picking up a calculator. This aligns with what children need as they grow. Handwriting improves foundational fine motor skills (Seo, 2018; Winslow, 2011). Visualizing numbers develops foundational numeracy (Geary, 2011). Then, they move on to more writing and mathematics. Students of any age might produce undergraduate-level essays with AI from a single prompt, but the output has little meaning without students' capacity to parse multilayered arguments, evaluate evidence, and connect that evidence from throughout the essay to the thesis and conclusion.

So, what *is* developmentally appropriate?

While there isn't yet a scholarly consensus on what elements of AI are best for each grade level, literature on child and adolescent development, combined with emerging recommendations from across the education sector, do suggest a curriculum that should be scientifically verified.

Elementary School: Lay the Foundation for AI as a Concept

At this stage, thinking is primarily concrete and tied to observable facts (Piaget, 1972). Working memory and self-control are still developing (Diamond, 2013), and moral reasoning focuses on concrete rules and consequences (Kohlberg, 1981). Independent AI use is likely premature at the elementary school level. Instead, teachers might create highly structured environments for students to use AI, with very specific goals, and only with well-tested tools. Or they might incorporate AI as a guided demonstration (for example, AI generating an image related to a story), which might offer students opportunities to connect technology to learning goals. In later elementary years, asking functional and ethical questions ("How do machines know things?" or "Did AI make that picture all by itself?") builds curiosity and foundational epistemological understanding.

Middle School: Build AI Awareness Through Guided Use

In middle school, abstract thinking emerges in familiar contexts, and students become more sensitive to peer influence and social dynamics (Chein, Albert, O'Brien, Uckert, & Steinberg, 2011; Eccles et al., 1993). Similarly, moral reasoning expands to include social approval (Kohlberg, 1981), and metacognitive skills begin to take shape, though self-monitoring and time management skills are still developing (Demetriou, Spanoudis, & Mouyi, 2011). At this stage, guided AI use might support students' brainstorming, translation, data analysis, or collaboration, with a high degree of teacher oversight and student metacognitive accountability. Teachers should pair students' AI use with lessons on its bias, reliability, and limitations, much like they do with media literacy. AI remains the exception rather than the norm in daily routines, and scaffolding student reflection is central to AI use.

High School: Integrate AI Into Learning With Critical Engagement

By high school, the necessary developmental foundations have formed for students to engage with AI more regularly in a wider variety of learning contexts. Students are capable of broader abstract reasoning and hypothesis testing (Piaget, 1972), though executive control may still falter under cognitive load (Casey, Getz, & Galvan, 2008). Moral reasoning now introduces the possibility of more complex ethical debate (Kohlberg, 1981). Students might engage with AI regularly for research, synthesis, and revision, as well as coding and data analysis. Teachers should emphasize AI literacy—prompt design, error detection, and cross-checking—and connect use to ethical considerations such as privacy, bias, and intellectual property. Students should be able to challenge and critique AI tools. Teacher guidance remains important for fostering disciplined habits, but AI use might become the norm in many contexts, balanced with non-AI work.

College and Early Adulthood: Expect Independent, Ethical, and Discipline-Specific AI Use

At the university level, students are prepared to move into self-authorship. In this period, executive functions reach maturity (Diamond, 2013), and students gain epistemological sophistication (Perry, 1999) and the potential to apply ethical reasoning consistently across contexts (Kohlberg, 1981). AI can now be fully integrated into workflows, from discipline-specific problem solving to creative production and simulation modeling. Students should take full ownership of their work with AI and should be able to explain and intentionally shape their AI use, applying it to extend disciplinary reasoning and deepen scholarly originality. College-level work should prepare students to make nuanced judgments about when to rely on AI and when to work without it, maintaining both personal agency and professional standards.

Across these levels, AI use will vary by discipline. Further, as both the technology and our practices evolve, the granular implementation will require teachers and leaders alike to make decisions.

For students, the progression follows natural developmental stages, and, in most cases, students will learn to perform skills on their own before using AI to help them. Like the entry-level lab assistant repeating experiments before rising to lead the lab, the summer associate proofing documents before taking on the position of legal counsel, and the line worker working the machines on the factory floor before becoming a manager, students accumulate wisdom and mastery through experience that then helps them when they can use tools. What does this look like in disciplines? In English, students might benefit from mastering essay structures before asking AI to assist. In science, they might design and interpret data from experiments before delegating those steps to AI. Learning by doing builds the deep understanding necessary to take the executive role.

There is a bolder school of thought, however, that points out that while executives typically rise from within an organization, thereby understanding how the organization works, they don't know how to perform many of the key responsibilities of the people who work for them. Likewise, the leader of a creative team typically can't do the work of the other team members. This bolder school of thought therefore asks, "Is it necessary for students to learn to write coherently on their own if they have the analytical ability to assess whether their AI assistant writes effectively on their behalf? If writing serves to help us think and communicate—and if we can judge whether AI-generated text adequately conveys our ideas—does personal mastery of the craft remain essential?" These questions will likely divide educators across philosophical lines for decades, if not centuries, to come. But they are not unreasonable. After all, most of us do not know how to hunt or farm the food we eat. Most of us do not know how to build the homes we live in or the digital devices we use every day. Does writing as a technology serve a fundamentally different purpose, and is that purpose undercut by AI? Perhaps it depends on our disposition.

Dispositions for Deeper Learning With AI

Will students hike, or will they teleport? In our age of AI, how much will they surrender their agency and become passive consumers, and how much will they take advantage of their new tools and act as executives of their own lives? These are questions not of knowledge and skills but of dispositions. The ways students approach learning, their attitudes toward challenge, and the values they bring to their work will shape their choices just as much as—if not more than—their abilities. We believe the following dispositions—curiosity, perseverance, collaboration, and ownership—determine whether AI becomes a shortcut or a catalyst for deeper learning.

Curiosity

Curiosity comes first. When curiosity is strong, we are intrinsically motivated to seek out new information, ideas, and perspectives. In an information-rich age, when well-formulated arguments and multiple perspectives are just a few clicks away, a student's self-driven wonder about the complexity and ambiguity of the world will help them not only most effectively harness AI's potential but also steer AI with their individual, human directions.

But the challenge isn't just wondering about and accessing all this information; it's about doing the work of integrating it meaningfully into mental models of the world. AI outputs require close reading and analysis. Content-heavy comparison tables and visual syntheses produced by AI reveal relationships among ideas, but interpreting these representations requires hard work. Even high school students often struggle to read tables well, sometimes more than they struggle with graphs (Sharma, 2006). Curiosity fuels the habits of actively engaging AI outputs and synthesizing them into students' own coherent understandings. It motivates the hard interpretive work of questioning evidence, detecting contradictions, and connecting new information to prior knowledge. Students (and many AI companies) seem to think that AI can do this sensemaking for us, offering and using services that address short-term informational needs while neglecting long-term learning (Bastani et al., 2025). But interpretive practice is essential for constructing knowledge from AI outputs and connecting it to existing knowledge structures in students' minds (Bransford et al., 2000), and curiosity is a powerful intrinsic motivator for developing this interpretive practice and building foundational knowledge.

The strong foundational subject knowledge that curiosity helps sustain is more important today than ever (Holmes, Bialik, & Fadel, 2019). Students who know their subject well can ask more precise questions with AI, more readily detect AI hallucinations, and assemble disparate information into richer insights. Further, the foundational knowledge they develop reduces cognitive load and supports more strategic help seeking (Dong, Jong, & King, 2020). Curiosity is the driver of this knowledge building. It sustains deeper and ongoing inquiry into subjects. It compels students not to be satisfied with initial answers but instead to search for contradictions, disconfirming evidence, and complexities that challenge their thinking.

Curiosity will also help inspire the perseverance to go the extra distance.

Perseverance

When necessary effort is low, expectations can be high. The internet enabled teachers to push students with more challenging research projects than in the days of card catalog and encyclopedia-based research. AI raises the bar further. If a first draft or a dataset can be produced instantly, the students can be—*ought to be*—expected to iterate, refine, and improve their work repeatedly. This requires a commitment not only from the teacher to set higher expectations but also from the students in what they expect from AI.

Today's learners have grown up in a world where algorithms deliver seamless answers without requiring them to click on sources or compare alternative perspectives. Those answers often arrive highlighted at the top of search results, requiring no more than asking once. But settling for first answers like these undermines the persistence needed to refine ideas and develop new ones. It reduces productive struggle. Using AI most effectively, however, demands a mindset of iterative experimentation: trying multiple prompts, analyzing where outputs fall short, and repeatedly adjusting inputs or evaluation criteria to improve results.

In this way, AI becomes less a shortcut than a tool for deep iteration. Students who learn to approach AI as a persistent collaborator like this—rather than as a vending machine with quick responses—will not only strengthen their capacity for sustained intellectual effort but also find the greatest value in their AI partnership.

And this partnership with AI, as the assistant-oriented structure of this book suggests, requires a commitment to collaboration.

Collaboration

One thing that's *not* going away in our AI world is humans collaborating with humans. Learning with AI might seem to undermine this, but learning experiences in which students can develop greater mastery on their own outside of class mean that our time together in class can focus more on our human-to-human relationships. As a result, students must develop strong collaborative skills—both with each other and with their AI tools.

Group work should be a natural setting for developing collaborative skills. At its best, group work exposes students to diverse viewpoints and enables them to produce outcomes richer than any individual could create alone. Yet, research suggests that collaboration in schools often falls short of this ideal, where students may learn less in groups than they do individually (Bacon, 2005), in part because classroom projects seldom include the clear roles, structures, and norms that support effective teamwork in professional settings.

Both poison and cure, AI has the potential to both undermine and enhance student-to-student collaboration. When used uncritically, it can reinforce isolation by compliantly catering to individual satisfaction rather than collective goals. (We described this briefly in chapter 4.) But if deliberately integrated for prosocial purposes, AI can help students develop the interpersonal skills necessary for successful collaboration, both with each other and with AI. For example, AI can mediate conflicts by analyzing disagreements from multiple perspectives, helping groups articulate shared norms, and suggesting constructive next steps. In one study, AI served as a mediator in students' chat conversations, detecting when the discussion was veering off track and intervening in ways that reduced interpersonal blame—functioning like a "moral crumple zone" (Hohenstein & Jung, 2020).

To make the most of these possibilities, students will need to practice collaboration in two intertwined arenas: (1) human-to-human collaboration and (2) human-to-AI collaboration. The first requires the empathy and conflict resolution skills that have always underpinned successful teamwork. The second demands the ability to integrate AI as a partner—one that can surface diverse perspectives, accelerate certain tasks, and even mediate disagreements but still requires human judgment to set goals and interpret outcomes. In an AI-infused world, the students who thrive will be those who can move fluidly between these modes of collaboration, harnessing technology to strengthen—not replace—the human connections that give learning its meaning.

And that collaboration will be possible when each participant takes full ownership of their role, their contributions, and the quality of their work.

Ownership

Ownership begins with choosing a direction. As we described earlier in this chapter, students are the executives of their AI teams. Effective executives don't just oversee the work; they decide what the work will be. They set the vision, frame the problem, identify what matters most, and define the moral and intellectual boundaries within which the work will proceed. For students, ownership means selecting a purpose they believe is worth pursuing, shaping their AI's role in achieving it, and committing to the process until completion.

Taking ownership, however, goes beyond choosing a direction; it also means taking accountability for the outcome. Just as in times before AI, students remain responsible for the quality and integrity of what they produce—no matter how much AI contributes along the way. This means students must critically assess AI-generated content, decide whether it meets the highest standard they can achieve, and make revisions that reflect their own thinking. It also means maintaining academic honesty, acknowledging the role AI played in the process, and ensuring that the final product represents their own best judgment. AI can generate words, images, data, and code, but it cannot take responsibility for their accuracy or meaning. That responsibility lies with the student.

In this way, ownership demands both setting a course that reflects one's own values and aspirations *and* having the commitment and courage to stand behind what has been produced. Just as a capable executive turns a team's diverse capabilities into a shared strategy, the student who takes ownership turns AI's capabilities into meaningful work—work they can stand behind because they chose its direction from the start.

Taste, Trust, and Teamwork: The Irreplaceable Traits

The skills outlined in this section—executive capacity, AI literacy, developmentally appropriate practice, and essential dispositions—are not just discrete abilities. They are connected by deeper, irreplaceable human traits that shape how students might best navigate an AI-infused world. We call these taste, trust, and teamwork.

Taste

Taste is the capacity to discern quality, purpose, and meaning. When AI can generate dozens of plausible options in mere seconds, students' roles shift from producer to executive editor, curator, and director. They must decide what is worth building, refining, or discarding and why. Far from reducing the need for judgment, AI multiplies the design decisions students face. Cultivating taste in writing, research, science, or design becomes an essential part of their education. Without it, students risk being overwhelmed by the volume of what AI produces rather than empowered by the choices they can make with AI.

Trust

In a landscape of information abundance, fact is almost undiscernible from fiction. And in a time of dubious authorship, ownership is easily questioned, and human accountability becomes increasingly important. Building trust means building confidence in the thoroughness and veracity of student work, and it means building confidence so that students personally stand by what they produce. This means not only critically evaluating AI-generated content for accuracy and bias to affirm its reliability but also sustaining attentiveness to whether using AI supports or erodes student ownership and direction. Further, it means standing behind and being accountable for the integrity of work within the academic context of school. At its deepest level, fostering trust is a moral and epistemological commitment to human agency and authority in an AI-saturated world.

Teamwork

Teamwork is the ability to strengthen human relationships through collaborative work. AI assistants are frictionless collaborators, but human relationships are full of misunderstandings and negotiation. Friction is not a bug in our human system, though—it is a feature, one that fuels creativity, insight, and understanding. As AI companions and assistants create the semblance of connection without the presence of other people, schools must protect spaces where students practice listening, disagreeing, cocreating, and navigating complexity together. In an AI-rich era, the most successful learners will be those who can move fluidly between working with machines and working with people, drawing on the strengths of each without losing the uniquely human capacities that give learning meaning.

Together, taste, trust, and teamwork are the connective tissue between the skills and dispositions students need to thrive. They are the compass points that can guide a student executive's decisions in an AI-rich world.

They can also be the compass points for what we need to do as teachers in an AI world.

Teachers' Needs

How must we concretely change our practices for a time when students are leading a team of creative AI assistants?

The Strategy: Rethink Learning Experiences

Let's return to the hiking metaphor. If an assignment focuses on reaching the summit of the mountain, then from the student's perspective, the only reason not to use AI and teleport to the top is the fear of getting in trouble. This kind of assignment and disciplinary model does not instill a love of learning; it instills a fear of breaking the rules.

But if we design with the teleportation machine in mind, hiking might look different. Perhaps, by teleporting from one navigational decision point to the next, students could spend less time walking and more time making navigational decisions. Or they could double back on the trail, revisit earlier intersections, and compare different routes—all of which wouldn't be practical if they were on foot, or even in a car. Of course, other times, they would want to choose not to use the teleportation device, perhaps instead walking with purpose from point A to point B, taking in the views, or watching for specific types of birds native to that particular area.

AI makes it both more important and more feasible to focus on higher-level executive and navigational skills. It's therefore up to us, as teachers, to make sure assignments keep up. Sometimes, the changes will be small, such as revised guidelines in our course expectations. Other times, we will need to entirely reimagine what we are doing. Here, we focus on five high-level strategies to achieve this new vision.

1. Focus on process over product.
2. Avoid and address pitfalls.
3. Raise expectations.
4. Trust autonomy.
5. Target student dispositions, not just deliverables.

These five strategies guide us to design student experiences that make shortcuts difficult and serve learning instead.

Focus on Process Over Product

Many teachers have been focusing on process over product for years. Strategies for doing this have included having students submit iterative drafts, requiring students to show their work, grading incremental steps, allowing revisions and retakes, and more. How might students' use of AI further support this kind of work, and how might teachers assess process-driven AI use?

The creative process (as explored in chapter 2) reveals the various stages of student work and, in turn, helps us guide how AI can support that work. In this section, we propose ways teachers might leverage AI to create new learning experiences. These approaches are organized around the stages of the creative process, with the exception of the first stage, incubation, which we explored in chapter 3, where we discussed how early feedback can expand the directions of student learning. Once this initial learning stage has taken place, students can be empowered to apply their learning creatively (Kalyuga et al., 2003), as in the following examples.

Experimentation

- *Require a best-of-three submission*—Students generate and compare at least three AI-assisted outputs before selecting one and documenting why they chose their final direction over the alternatives.
- *Design stakeholder simulations*—Students use AI to model different perspectives on their topic, then navigate the tensions among them to reach decisions that acknowledge trade-offs.

- *Implement failure portfolios*—Students document dead ends, rejected drafts, and pivot moments in their work with AI, reflecting on what each "wrong turn" taught them about their question or approach.
- *Design constraint challenges*—Students work with increasingly difficult limitations (such as to use only ten prompts, work with a specific AI tool, or combine three incompatible sources) to build resilience and creative problem solving.

Perspiration

- *Require source archaeology*—When AI makes a claim, students trace it back through multiple sources, documenting the "telephone game" of information transformation.
- *Create AI translation tasks*—Students take AI's output and translate it into forms the AI cannot produce (such as physical models, performed speeches, or visual narratives), requiring deep understanding.
- *Build remix assignments*—Students take each other's AI outputs and transform them for different audiences, purposes, or mediums to show how human judgment shapes raw material.
- *Run real-time workshops*—Students work on AI-assisted tasks live in class, like pair-programming in computer science, narrating their choices aloud while peers observe and offer alternative approaches.

Revision

- *Establish peer-review protocols*—Students exchange their AI conversation histories (not just final products) and critique each other's prompting strategies, decision points, and missed opportunities.
- *Create AI debate rounds*—Students use AI to generate opposing viewpoints on their topic, then defend their chosen position while acknowledging the complexity revealed by multiple perspectives.

Integration

- *Institute weekly reflections*—Students produce short video or audio recordings where they explain their biggest AI-related decision that week and whether they'd choose differently now.
- *Assess decision logs*—Students create decision logs that show prompt iterations and reflections, making their thinking visible through screencasts and "captain's logs." These are graded as heavily as the final outputs.
- *Design for authentic stakes*—Students choose real audiences and meaningful problems, rather than hypothetical exercises, so they experience the weight of actual decision making and learn to handle the responsibility that comes with directing powerful tools.
- *Assess disposition development*—Students explain their interactions with AI, show visible iteration in their work, expand project scope ambitiously, and maintain a personal voice, even within AI-assisted scaffolds.
- *Build mandatory power-off zones*—Students complete assignments with AI-free segments so that they must demonstrate core skills without assistance, which ensures they develop genuine capability alongside their orchestration abilities.

This list is just a beginning. Think about creative projects you assign and how you could incorporate AI at key moments to help students focus on the creative aspects of the projects rather than spend time and effort getting caught up in mechanics that are incidental to the intended learning goals.

Avoid and Address Pitfalls

While engaging students in AI-supported work, teachers might observe certain pitfalls and traps that students might fall into and apply "patches" from the previous list:

- **The First-Answer Trap**
  - *Pitfall*—Students uncritically accept the first answer AI provides.
  - *Patch*—Require a *best-of-three submission* and focus student attention on how to compare outputs.
- **Iteration Without Learning**
  - *Pitfall*—Students work through iterations as assigned but do not engage metacognitively.
  - *Patch*—Require a *decision log* and provide feedback on students' decisions throughout the process. Assess the decision log as heavily as the final product (or even more heavily!).
- **Voice Erosion**
  - *Pitfall*—As students use more AI, their work reverts to more generic AI patterns.
  - *Instructional patch*—Ask for work that AI could not have written, such as anecdotes, opinions, sensory details, or local knowledge. Provide students with AI-generated answers to the assignments for comparison.
  - *Cultural patch*—Create a class culture where you expect students to defend their choices, down to the word choice. If students begin using words that don't sound like them, ask them to defend those choices or make new choices.
- **Struggle Avoidance**
  - *Pitfall*—Students begin deferring more and more to AI for important steps in the work process.
  - *Patch*—Build mandatory *power-off zones* where AI is not permitted and core skills carry the weight.

These can, of course, be combined to put multiple patches to use in one intervention. For example, teachers might ask students to brainstorm relevant personal experiences without AI, then iterate on a write-up of the personal experience while providing a decision log as part of their final submission.

Raise Expectations

Just as we would expect a student with a bicycle to travel longer distances than a student on foot, so should we have higher expectations for students with AI. When required effort drops, ambition should rise.

If we incorporate ways for students to use AI, then assignments can ask students to achieve more ambitious and authentic goals. For example, when Maya taught middle school science, she emphasized experiment design throughout the school year. Toward the end of the school year, she picked out a scientific paper that was based on the science her students had learned and allowed them to use AI to analyze the independent, dependent, and controlled variables and identify confounding variables. Table 5.1 shows this example and similar examples from other subjects.

Table 5.1: *Raising Expectations by Redesigning Tasks to Use AI*

Subject	Traditional Task	AI-Supported Task
Science	Complete a test on the experimental design of in-class demonstrations.	Use AI to simplify a scientific paper selected by the teacher, analyze its experimental design, and suggest confounding variables.
History and Social Studies	Write a four-to-five-page essay on a historical controversy using teacher-supplied sources.	Work in teams to create a four-to-six-minute documentary-style video contrasting three or more viewpoints on one event.
English Language Arts	Compose a five-paragraph literary-analysis essay.	Draft a five-paragraph essay, then iterate on it with AI feedback. Require a reflection on the iteration process.
World Languages	Complete a two-minute oral routine, with slides.	Curate a bilingual 360-degree photo tour (such as on a morning in Quito).
Visual Arts and Design	Create a poster using one design principle.	Create an interactive art exhibit focused on a single design principle with a specific audience in mind.

Trust Autonomy

It isn't a coincidence that trust is a cornerstone for students and teachers alike. With greater confidence in the foundational mechanics of student work, teachers can provide greater latitude in what they assign to students, thereby allowing them the opportunity to take greater ownership and practice agency. In the following examples, students might make decisions about topics, stakes, and audiences to make the experience more self-driven and authentically engaging.

- **History and Social Studies**
    - *Core learning objective*—Students analyze and evaluate the reliability, perspective, and historical context of a primary source.
    - *Autonomy*—Students choose a lens (such as a journalist or activist) and a format for response.
    - *AI support*—Students ask AI to simulate alternative interpretations, generate historical context, or reframe documents through different ideological filters.
    - *Key student work*—Students engage in a *stakeholder simulation*, using AI to generate multiple ideological or cultural interpretations of the same source and negotiating which interpretation to advance and why.
    - *Real stakes*—Students publish a response letter or record a mock interview.
    - *Potential grading criteria*—Students may be graded on insightfulness, contextual accuracy, originality, and thoughtful AI use.
- **English Language Arts**
    - *Core learning objective*—Students interpret literary symbols and analyze how they develop meaning within a text.
    - *Autonomy*—Students choose how to represent their interpretation, such as through an analytical paragraph, a visual metaphor, a short film, or an AI-powered chatbot.

    - *AI support*—Students ask AI to brainstorm symbolic interpretations, generate alternative formats or narrative framings, and model potential meanings.
    - *Key student work*—Students maintain a *failure portfolio* documenting misinterpretations, symbolic red herrings, and AI-generated misreadings that sharpened their final claim.
    - *Real stakes*—Students contribute to a class digital gallery or submit to the school's literary magazine.
    - *Potential grading criteria*—Students may be graded on depth of interpretation, use of evidence, intentional design, and ethical use of AI.
- **Mathematics**
    - *Core learning objective*—Students construct and interpret linear models from real-world data.
    - *Autonomy*—Students choose personal or societal datasets and analyze trends with linear models.
    - *AI support*—Students ask AI to generate or clean data, identify correlations, draft equation structures, and assist in error analysis.
    - *Key student work*—Students complete a *source archaeology* task: When AI proposes a correlation or regression insight, students trace its logic backward through data sources and critique the appropriateness of the model.
    - *Real stakes*—Students present in a classroom data gallery or submit to a school publication.
    - *Potential grading criteria*—Students may be graded on accurate modeling, relevance of data, thoroughness of critique, and transparency about AI use.
- **Computer Science**
    - *Core learning objective*—Students implement and compare basic sorting algorithms and explain their computational efficiency.

- *Autonomy*—Students choose a scenario, apply multiple algorithms, and explain trade-offs.
- *AI support*—Students ask AI to offer code scaffolds, debug snippets, and visualize sorting processes.
- *Key student work*—Students participate in a *power-off zone* where they must explain, by hand and orally, the logic of each algorithm after having iterated their code with AI's help, showing they own the thinking, not just the result.
- *Real stakes*—Students create a peer-targeted tutorial with a reflective appendix documenting AI versus human contributions.
- *Potential grading criteria*—Students may be graded on correctness of implementation, clarity of explanation, insight into trade-offs, and independent reasoning.

In each case, AI serves as a support for learning, and students have greater autonomy and freedom to explore. Further, the real stakes involved in each invite students to take ownership of what they produce in concert with AI.

Target Student Dispositions, Not Just Deliverables

By focusing on process rather than product, learning experiences draw greater attention to student dispositions. Curiosity will outlast the rubric for any one assignment. Perseverance and iteration, ownership of their submitted work, the habit of collaboration—these dispositions similarly lead to student success one way or another. Curriculum permitting, cultivating these dispositions might be a new gold standard at your school.

Student dispositions are already the focus of many school-improvement efforts, especially those working on a Portrait of a Graduate. But this less concrete work is often challenging to implement, as discussions of dispositions tend to be abstract. Now, AI can be an impetus and mechanism for bringing dispositions into the everyday work of learning.

Rather than a vague hope for our students, these dispositions can be explicitly named as learning goals, with evidence of learning coming from students' interactions with AI. How many questions did they ask? What kinds of questions? How many follow-up questions? How many interactions did they persevere through? What did they accept, and what did they add in themselves?

AI in Practice: Suggested Signs of Effective Use

If students are using AI as a support for their learning, developing agency while using AI, and learning both how to hike and how to navigate, then we will see signs like the following.

- Students can explain *why* they prompted, revised, or ignored AI.
- Student drafts show visible iteration and elevation.
- Project scope expands to include bigger datasets, real audiences, and interdisciplinary twists.
- Individual student voice emerges, even inside AI-generated scaffolds.

The strategies in this section are less about integrating a new tool and more about reshaping the conditions in which learning happens. By focusing on process, addressing pitfalls, raising expectations, trusting autonomy, and targeting dispositions, we create classrooms where AI amplifies rather than replaces student thinking. But even the most thoughtful design can falter if students are unsure about when and how AI is appropriate. That's why an important step is to make your expectations explicit by establishing clear, transparent guidelines that help students use AI with purpose and confidence.

The Tactics: Communicate Clear AI Use Guidelines

In a 2025 survey, 78 percent of educators reported being "very" or "somewhat" worried about students cheating with AI (Carnegie Learning, 2025). The same survey found that only 40 percent of schools and districts have an AI policy in place (Carnegie Learning,

2025). Teachers have an urgent need to tell their students *something* about how they can and cannot use AI.

The AI Assessment Scale

One widely used framework is the AI Assessment Scale by Mike Perkins, Leon Furze, Jasper Roe, and Jason MacVaugh (2024). What is immediately useful about this framework is that it does not simply cast aside all AI as cheating, and it is a simple, usable progression that teachers can add to their assignments as soon as they see fit. This scale presents AI use on a linear trajectory, from level 1 through level 5.

1. **No AI:** Students complete the assessment entirely without AI assistance in a controlled environment to ensure that they rely solely on their existing knowledge, understanding, and skills. They must not use AI at any point during the assessment. They must demonstrate their core skills and knowledge.
2. **AI planning:** Students may use AI for pretask activities such as brainstorming, outlining, and initial research. This level focuses on students' effective use of AI for planning, synthesis, and ideation, but assessments should emphasize students' ability to develop and refine their ideas independently. They may use AI for planning, idea development, and research, but their final submissions should show how they developed and refined their ideas.
3. **AI collaboration:** Students may use AI to help complete the task, including idea generation, drafting, feedback, and refinement. Students should critically evaluate and modify AI's suggested outputs to demonstrate their understanding. They may use AI to assist them with specific tasks, such as drafting text and refining and evaluating their work. They must critically evaluate and modify any AI-generated content they use.
4. **Full AI:** Students may use AI to complete any elements of the task, directing AI in order to achieve the assessment goals. Assessments at this level may also require students' engagement with AI to achieve goals and solve problems.

They may use AI extensively throughout their work, either as they wish or as specifically directed in the assessment. Students' focus is on directing AI in order to achieve their goals while demonstrating their critical thinking.

5. **AI exploration:** Students may use AI creatively to enhance problem solving, generate novel insights, or develop innovative solutions to problems. Students and educators codesign assessments to explore unique AI applications within the field of study. They should use AI creatively to solve the task, potentially codesigning new approaches with their instructor.

The authors of the AI Assessment Scale present it as a much-needed starting point for empowering teachers to think about how to move beyond the discourse of student use of AI as academic dishonesty.

If this is the starting point, where should we go from here? We present four possible directions and our attempt to contribute an updated framework that achieves these four requirements.

1. **From cumulative scale to checklist:** The AI Assessment Scale presents different uses of AI in a scale that appears linear, increasing, and cumulative. As they move up the levels, the scale suggests that students do less of the work and AI does more of it, so when a teacher indicates that an assignment permits level 3, it also permits levels 2 and 1. Some teachers, however, may want to empower students to use AI for *refinement* (level 3) but not for *outlining* (level 2) so that they practice the executive skill of framing the problem on their own while using AI to push their writing mechanics further than they could without assistance. *AI use could be represented in a checklist of discrete tasks that teachers select that align with the learning goals of the assignment, rather than a linear scale.*
2. **From AI centric to student centric:** The AI Assessment Scale frames AI use around what AI can do rather than what using AI allows *students* to practice doing. For example, level 3 (AI collaboration) describes a range of AI capacities, including

drafting text and refining and evaluating student work. Since our goal is student learning, we propose shifting the focus from AI to the students. *Acceptable uses could be framed through a focus on student actions or learning objectives rather than degrees of AI functionality.*

3. **From general use case to specific use case:** Elements of the AI Assessment Scale describe general, overlapping use cases. This is conceptually inclusive; however, it can also lead to confusion. For example, level 3 invites AI use to help complete the task, whereas level 4 welcomes AI use to complete any elements of the task. This may be a distinction that requires additional clarification to resolve. *Enumerating different, more specific uses of AI may offer more precise clarity than more general use-case descriptions.*
4. **From assessment focused to learning focused:** The AI Assessment Scale focuses on assessment, which omits other AI use cases for learning. For example, students might use AI to revise an assigned text to match their reading level or interests, provide motivation to engage more deeply with a task at hand, or create low-stakes practice opportunities beyond those provided by the teacher. These are powerful uses of AI that students may seek out—and teachers may want to invite—that would benefit from being clearly permitted or prohibited. *To broadly empower teachers and students, a framework could be more comprehensive while remaining intuitive and clear.*

The AI Student Use Framework

To try to achieve these four aims, we share the AI Student Use Framework (see table 5.2). We sought to balance clarity for teachers with greater inclusiveness of learning activities and more overt encouragement of productive uses of AI for learning. We aim for this framework to be as applicable to completing a project as it is to writing an essay or preparing for an assessment.

The framework is organized into AI uses before, during, and after drafts of a project or assessment. The student use column summarizes how a student might use AI. The learning goal column focuses attention on the purpose of this use of AI—to practice higher-order skills—and each goal includes a sample prompt which teachers or students can adapt to their needs.

Table 5.2: AI Student Use Framework

Student Use of AI	Learning Goal and Example Prompt for AI
Before the Project or Assessment (Building Necessary Knowledge)	
Motivation	**Learning goal:** Students practice expressing themselves and their passions. **Example prompt for AI:** "Explain why I should care about slope using Roblox."
Access	**Learning goal:** Students practice advocating for their needs. **Example prompt for AI:** "Rewrite this passage at a fifth-grade reading level. Define [words]."
Practice	**Learning goal:** Students practice focusing on mastery over performance with low-stakes practice. **Example prompt for AI:** "Generate a practice quiz with an answer key and provide explanations based on my notes." [Upload notes.]
During the Project or Assessment (Creating the Artifact)	
Brainstorming	**Learning goal:** Students practice choosing among a wide set of options according to their vision. **Example prompt for AI:** "Generate ten possible essay topics that would work for this assignment." [Upload assignment.]
Research	**Learning goal:** Students practice evaluating relevance of sources and integrating perspectives. **Example prompt for AI:** "I am working on a project about [topic] and my thesis is [thesis]. What are some sources that provide evidence for and challenge my point of view? Include links." (Make sure AI has internet access.)
Outlining	**Learning goal:** Students practice considering multiple structures and making decisions from different strategies. **Example prompt for AI:** "I am writing a paper on [topic]. Offer three possible outlines for this paper."
Planning	**Learning goal:** Students practice metacognitively planning out a project, sticking to a timeline, and reflecting on time management. **Example prompt for AI:** "I am doing a project in which I have to make a documentary about [topic], and it's due in fourteen days. Give me three options for a day-by-day plan of deliverables."

continued →

Facilitation	**Learning goal:** Students practice articulating reasons, comparing options, discussing as a group, and including everyone's perspectives. **Example prompt for AI:** "In a group assignment with these instructions, one person thinks we should do [option] because [reason], and another thinks we should do [option] because [reason]. What are the pros and cons of each? What are three possible compromises?" [Upload instructions.]
Drafting	**Learning goal:** Students practice articulating their goals and plans, seeing how different phrasings affect outputs. **Example prompt for AI:** "Write three versions of this paragraph given this outline." [Upload outline.]
After Completing a Draft of the Project or Assessment (Refining and Reflecting)	
Feedback	**Learning goal:** Students practice revising their work and writing at a higher level than they could alone. **Example prompt for AI:** "Here is my assignment, my paper, and the rubric I will be graded on. Suggest how I can improve my paper without rewriting any parts of it for me." [Upload assignment, paper, and rubric.]
Reflection	**Learning goal:** Students practice reflecting on their learning process. **Example prompt for AI:** "Here is my chat with AI while working on this project. What do you notice? Ask me some questions to help me reflect on my learning process." [Upload previous chat with AI.]

To optimize students' learning experiences, teachers might align their assignments with specific use cases, or they might indicate that some assignments may not use AI for any use cases. Teachers who want to empower their students with even greater *meta agency* (agency about their agency) might even allow students to choose a number of use cases—such as three to five—that they can use for a given project. From there, students can decide for themselves and, of course, be ready to defend their choices.

Use of an AI use framework like this one or the AI Assessment Scale must be accompanied by AI literacy training, intentional learning design for AI, and a classroom culture with clear expectations for how students use AI. With these components in place, ongoing AI schoolwork might lead to new academic habits in which students routinely:

- Share their chat logs
- Provide their reasoning for their choices during their chats
- Provide a reflection on how their use of AI affected their learning
- Show iterations (not just the final draft)
- Ask for multiple options so they can choose among them
- Ask for sources and check the sources themselves

Strategic use of AI can lead to deeper learning experiences that allow students to practice the higher-order skills and dispositions they will need for the future and would not have been able to practice without AI.

Writing in a Time of AI

It is worth taking some time just for the question of what happens to writing in an age of AI. Since the arrival of ChatGPT, many pundits have pronounced the essay dead (Hsu, 2025; Marche, 2022). They are mistaken, but they are looking in the direction of a truth: Writing's preeminence as a means of learning and communication is waning—not dying, but waning. The arrival of pervasive visual and auditory media, primarily consumed and increasingly produced with mobile devices, has upended how we interact and connect with each other. This merits mourning, for while we are gaining much, we are losing much, too.

Still, one technology does not always obviate its precursor. When the microwave was invented, it seemed the era of the oven might have ended. Who would bake something when it could be microwaved in a fraction of the time? But it turns out that food cooks differently in the oven than in the microwave. So now we cook some food in the microwave, some in the oven, some in the toaster, and even some in the air fryer. We use different technologies for different needs.

So it is with writing. Sometimes, we write by hand. Sometimes, we write with a computer. And sometimes, we write with both—just like we sometimes use the microwave to melt butter that we then apply to a turkey baking in the oven.

The Reasons Why We Write

So if different technologies serve different needs, what needs does the *technology* of writing itself serve? And how does AI change how we meet those needs?

The earliest forms of writing were accounting records. Mesopotamians in the 4th millennium BCE tallied inventory and transactions on clay tablets. The purpose of writing was to represent and communicate a fact in the world. Over time, writing expanded from numerical systems to pictographs and words, and we have used these forms to represent and describe more and more of the world. Much of our writing in schools today continues that tradition: the expository essay, the compare-and-contrast essay, and the argumentative essay. These represent information or ideas in the world. Similarly, datasets, computer code, lab reports, equations, and proofs—these also are forms of writing that show us the world. Together, essays, reports, proofs, and code reveal the first reason we write: to *communicate*. We write to convey thoughts and facts, and the discipline and skill of practicing this has served us well for centuries.

But writing has a second, often overlooked purpose. When we journal, take notes, freewrite, outline, reflect, map concepts, or annotate, we are writing not to communicate but to *understand*. Most writing doesn't begin in expository form, after all; it starts with marginalia, ideas, brainstorms, or other practices that seek to explore ideas and thoughts. Writer and researcher John Warner (2025) argues that writing *is* thinking and *is* feeling. In his book *More Than Words*, he makes the case that writing is the act of channeling our thoughts and feelings and putting them into a new form that helps us see and feel more of who we are (Warner, 2025). Writing to understand helps us become more self-aware. It deepens our humanity.

The Role of AI in Writing

Our two reasons to write—coupled with our teaching philosophies—help clarify whether or how we might use AI in writing.

Since AI composes clear, lucid prose, it is immediately supportive of conveying meaning. So, if our goal is to develop students who can leverage the tools around them to communicate efficiently and effectively, then AI ought to be as obvious a learning tool as pen and paper. If your teaching philosophy (from the introduction) leans toward pragmatic, then this seems a natural step; writing with AI will enable students to participate more fully in work environments, regardless of their personal ability to write unaided by AI. They will still need to review, revise, approve, and own what they write with AI, but AI is a natural tool.

It may seem that when writing to understand, the opposite is true. But surprisingly, if the purpose of writing is to understand, leveraging AI as a discovery partner might still be a natural step. With AI, students can explore ideas, test their thinking, and iterate quickly and easily in note, outline, prose, or other forms. AI also helps with understanding. So, if your teaching philosophy leans toward academic, then using AI for writing follows logically—for while having students use AI to understand and develop ideas does decenter professorial expertise, it nonetheless accelerates students' expression and mastery of concepts.

There is a caveat, however. If your teaching philosophy leans toward humanistic or social, you may feel that the purpose of education is more about self-discovery and finding and developing students' individual and cultural linguistic voices. And since writing produced by AI follows statistical patterns, it tends to homogenize or erase individual idiosyncrasies. If the purpose of writing in the classroom is to develop not just students' individual ideas but their individual linguistic voices specifically, then it follows that they should use AI sparingly, especially in the early and later stages of writing. This narrow focus on language may apply to only a subset of classes in which writing takes place, but it is worth highlighting.

The Right Tools at the Right Time

In the cooking metaphor, writing is heating something up. Sometimes, we want to slowly cook a turkey in the oven with sustained heat that develops a perfectly crisp exterior. Other times, we simply need to soften overly hard ice cream in the microwave.

And in the earlier metaphor of transportation, writing is travel. Sometimes, the distance is short, and walking or a bicycle will do.

Other times, if we could teleport, we would. Still other times, we may choose to bike the long distance because the exercise makes us stronger or reduces our impact on the environment. Some people commute by bicycle, even long distances.

So it is with AI. Sometimes, we work without it to experientially understand what it takes to construct a sentence, paragraph, or essay. Other times, we may simply need a clearly communicated statement, and a simple prompt will do. Still other times, we will work iteratively, both writing on our own and leveraging AI to create, review, revise, and more. Our objectives will determine which approach we take at which times and why.

Review and Next Steps

Student access to AI—to tools that now support not only cognitive tasks but, increasingly, aspects of social and emotional engagement—marks a turning point in how students work and learn. In this moment of change, our instinct may be to grip tightly to what is being lost: traditional forms of effort, expression, and mastery. But preservation without exploration belies a belief that we can stop time. This is a mistake, but it is one born of legitimate grief and of some truth: There are past practices that are worth not only holding on to but elevating. Instead of freezing at this time of transition, we must return to the higher-level goals we have for our students and then revisit what we know, separating what is essential for achieving these goals from what is merely familiar to us as teachers. Many of the enduring understandings from the past will—and must—serve as the foundation for what comes next.

At the same time, we must look ahead to recognize and understand the new capabilities that AI enables. Rather than ignoring or rejecting them, we can ask how these new capabilities might advance our priorities and raise our expectations. Then, from this dual effort—discerning what must remain from the past and embracing what is newly possible—emerge the human traits that are newly relevant and essential in an AI-saturated world. We believe these traits are taste, trust, and teamwork and all that they contain. From our study, these are the three irreplaceable human characteristics that our learning experiences can and should be developing in our AI age.

In a time when AI's capabilities continue to expand, the question facing education is no longer whether students can complete a task to demonstrate their skills on their own but whether they can leverage their skills and tools to set and complete their own tasks. If we do not teach students to define their own direction, technology will continue to define more and more of it for them. That is why taste, trust, and teamwork are not optional enhancements; they are foundational capacities for human agency in an AI-saturated world. These are the traits that will allow students to frame meaningful questions, evaluate complex possibilities, and build lives and communities with others—not just efficient lives but ethical, joyful, and independent ones. As educators in this changing world, we must ensure that teaching and learning with AI expand—rather than erode—these human capacities at the center of learning and life.

What follows are next steps that we believe could help move us toward that positive alternative future.

For Educators

The following are some considerations for educators related to AI for student work.

Evaluate your assignment's learning objectives. Choose an assignment that you'd like to revise in the context of AI. Consider which learning objectives you want to use the project as a practice ground for. Once you've listed out your learning objectives, imagine a student who completed all learning objectives themselves but used AI for everything else. Would that be OK? If not, add the missing learning objectives to your list.

Redesign the assignment through the lens of the creative process. Imagine the specific steps your students take to complete an assignment and organize the work students do into the different stages of the creative process. Then, for each stage, consider how students might use AI to augment their thinking in two ways: (1) by scaffolding the learning objective and (2) by adopting nonessential tasks. Last, consider what guardrails might prevent students from using AI to avoid thinking. As you take these steps, consider brainstorming each of them with your students. The same way you might set norms for your classroom, set norms for AI use in an assignment. Consider setting norms collaboratively with your students.

Talk with your students about AI. Even decades into the future, when AI is pervasive in school and life, students will need to be introduced to AI just as they are to the use of calculators and other technologies. AI literacy will be another element of media literacy, and we will need to help students develop habits for responsible usage. Then, the work of developing this literacy will be scaffolded throughout school curricula. Now, however, it is all new, and it is a big shift for both students and teachers. The only way to identify and begin practicing the literacies that are best for each school and classroom is to talk with our students about AI: how it works, what its risks are, and what responsible usage looks like. Help them understand it metacognitively. How does it feel to offload thinking to AI versus using AI to scaffold or support thinking? How can we be discerning about when we want to use AI?

It will take time for society to develop healthy habits and norms around AI use. Just as with plagiarism, it will be necessary to explicitly talk about what academic integrity violations with AI look like, what makes them violations, and how to avoid them in the future. And, just as with the calculator and other technologies, it will be necessary to model, discuss, and continually update what healthy and effective practices look like and how to set the conditions for future success.

Don't use AI detectors. It may be tempting to solve the problem of students using AI to cheat by using AI detectors. However, mounting evidence shows that these tools are deeply flawed—not just technically but ethically. Studies have found that AI detectors are inherently unreliable and risk penalizing students unfairly due to false positives, even when AI involvement is minor (Dalalah & Dalalah, 2023; Elkhatat, Elsaid, & Almeer, 2023; Weber-Wulff et al., 2023). Equally troubling is the bias against the writing of non-native English speakers that was revealed in a Stanford-affiliated study: Over

61 percent of essays by non-native English speakers were misclassified by AI detectors as AI generated, far higher than essays by native English speakers, with nearly 20 percent of cases unanimously flagged across AI detectors (Liang, Yuksekgonul, Mao, Wu, & Zou, 2023). These false accusations can damage reputations, provoke stress, and undermine trust—especially for students already navigating language barriers.

That said, the challenge of preserving academic integrity in our age of AI is real. Institutions are right to want fair, reliable mechanisms to detect or deter misuse. But integrity is better preserved through trust, clear guidance, and thoughtful assessment design than through a reliance on error-prone tools. Educational research consistently references clear communication of policies, open dialogue about appropriate and inappropriate use, and assignments that make process visible—such as iterative drafting, in-class reflections, and other approaches, like those in this chapter—as paths for integrating AI confidently and responsibly (An, Yu, & James, 2025; McDonald, Johri, Ali, & Collier, 2025; Tillmanns et al., 2025). These pedagogical approaches reinforce integrity through clear articulation of goals and communication of objectives. Moreover, involving students in conversations about AI and redesigning learning experiences to reduce incentives for misuse can be more effective than using detectors that too often disproportionately punish the vulnerable.

For School Leaders

The following are some considerations for school leaders related to AI for student work.

Develop clear guidelines for appropriate use of AI. It's tempting to focus on the potentially disastrous effect that AI could have on student learning. After all, its application to cheating is obvious and inescapable. But before banning it completely, remember that (1) banning it will not stop students from cheating (they are already cheating!), and (2) you don't want to throw the baby out with the bathwater. There are many ways that students can use AI—just as adults can use AI—that will train them to think more deeply, openly, and compassionately than they could without it.

For better and for worse, addressing the issue of students cheating with AI is going to be more complicated than adding another rule to the student handbook under the Academic Integrity section—it will require teachers to engage deeply with their assignments and then with their students. It will take devoted efforts to clarify and practice. Luckily, teachers are already quite worried about this, and thus invested in fixing the problem! Examine your school's Portrait of a Graduate (or equivalent) and consider how this graduate is affected by living in a world with AI. Review the principles in this chapter and come up with your own, where relevant. Agree as a school on the vision for how the school prepares students for the AI future. From that, extract the learning outcomes for responsible AI use that students will need to practice throughout their time at the school.

Build engagement with AI use guidelines into your staff professional development schedule. Simply deciding on learning outcomes, even if teachers are involved in the process, is not going to be enough. Teachers, like all learners, learn by doing. They will need dedicated time built into the schedule to work together to wrap their minds around how AI guidelines apply to their classroom, what learning objectives they want to highlight in which assignments, how they might revise assignments, and how to talk with students about AI. Give teachers this space (or give up and do not expect change) because this will take effort on their part, and it cannot be an afterthought.

For Technologists

The following are some considerations for those working in the technology industry related to AI for student work.

Resist the instinct to remove all friction. Productive struggle is necessary for learning. Resist the temptation to solve students' problems as you might when designing a consumer app. Remember that solving a student's problem denies them their learning. Instead, think about the different ways AI might be helpful in removing extraneous work or scaffolding intrinsic cognitive load in the learning process. Think about the same learning objectives as the teachers, and design tools that develop the long-term skills students will need in a world of AI.

Design for partnership with the teacher. Student use of AI as a doing assistant should be prosocial. How does the tool you are building strengthen the relationship the student has not only with the course content but the teacher? The tool should drive conversation with a teacher, not replace it. Meyer (2025) argues that minimizing a teacher's role to making sure students log in to the app is a waste of their expertise, whereas a true partnership would allow teachers and AI to achieve more together than they could on their own. AI and humans working together have an advantage over AI alone or humans alone, as has often been demonstrated in other contexts already, such as in medicine (Zöller et al., 2025).

Even in finance, one study used AI to predict stocks and compared the results to human predictions. AI outperformed humans in predictions involving transparent and structured data, but humans outperformed AI when "human expertise and institutional knowledge play a critical role," such as "in the presence of intangibles" and when there are "rapid changes" or "higher distress" (Cao, Jiang, Wang, & Yang, 2024). When they were able to effectively work together, however, the AI-plus-human teams outperformed either one alone. However, this collaboration is not automatic or necessarily easy. Research suggests that it is more effective to design technology to be easily integrated into human collaboration patterns than to optimize for accuracy alone (Bansal, Nushi, Kamar, Horvitz, & Weld, 2021). The lesson for technology designers is that if AI is to support student and teacher success, it must be built to integrate with human systems and patterns.

Part 3

AI FOR THE CLASSROOM

No plan survives contact with the enemy.

—CORRELLI BARNETT, REPHRASING HELMUTH VON MOLTKE

Mathematicians estimate that there are somewhere between 10^{111} and 10^{123} possible chess games (Shannon, 1950). 10^{111} is an incomprehensibly large number. Over a billion times larger than a googol (10^{100}), it looks like this: 1,000,000,000,000,000,000,000,000,000,000,000,00 0,000,000,000,000,000,000,000,000,000,000,000,000,000,000,000 ,000,000,000,000,000,000,000,000,000,000. This number is larger than the estimated number of atoms in the universe (Baker, 2021).

We estimate that there are even more potential classroom experiences.

While every chess game begins with thirty-two chess pieces on the board that always start in the same orderly positions, every class begins with a room full of often-not-orderly students, each different than they were the day before.

Maybe a student didn't have breakfast that morning, or their significant other broke up with them that day, or they had a challenging assessment in the previous class, or a local or global crisis has disrupted their mood. Regardless of the cause, one of a teacher's most important objectives in the first five minutes of every class is to bring everyone together to a common starting point. Even then, once the learning experience has begun, students attend to different details, manage their own emotional health, activate varied prior knowledge, navigate their constantly evolving social impulses, and, when everything works right, engage with the learning experience.

Now imagine that every piece on the chess board—every pawn, rook, knight, bishop, king, and queen—has its own perspective on what's happening during the game. And imagine that, at every moment, each chess piece is choosing whether and how much to follow the direction of you, the chess player. Now, in addition to determining the next best chess move to win the game, you also need to keep every chess piece focused on the game, engaged in your strategy, and willing to follow your direction. Sometimes, you need to figure out how to bring that pawn back to the chess board, clear space for that knight that moves differently from everyone else, and convince that bishop that it should move diagonally because, for reasons unknown to you, it has chosen to move horizontally.

Teachers don't only set and follow a strategy; they also coach and cajole each chess piece. In the classroom, the dynamics of each piece are as much (if not more) of a challenge than the strategy of the game itself.

In part 3, we explore the complexity of the classroom, and we position the teacher as orchestrator of the administrative and pedagogical experiences of the classroom. The next two chapters explore an AI assistant that supports each of these two functions.

- **Chapter 6, "The Administrative Assistant":** Before, during, and after learning takes place in class, a teacher manages an extraordinary amount of logistics: attendance, technology, physical materials, notes from class, and more. Chapter 6 focuses on how AI tools assist teachers in the classroom setting. Specifically, we look at teachers' cognitive load and invisible labor and how AI can support more of the logistical and administrative work of the classroom so the teacher can focus more on students.
- **Chapter 7, "The Teaching Assistant":** Once learning begins, a teacher not only executes a plan but also actively manages the shifting dynamics of the students in the classroom. A teacher

who is masterful in pedagogy will be attentive and responsive to many changing student needs, and they will vary their instructional strategy based on the signals coming from the students. This is the focus of chapter 7: how teachers attend and respond to the evolving experience of the classroom and how generative AI can, like a team teacher, empower teachers and students with the right information at the right time, making class even stronger.

CHAPTER 6

The Administrative Assistant

A Brain Extension for the Invisible Labor of Mass Learning

Teaching isn't rocket science. It's harder.

—RYAN FULLER

The human brain can only do so much. From infancy, we learn what to pay attention to and how to ignore the rest. Without this filtering process, the world is a nonstop, dizzying onslaught of sensory perception.

For teachers, it is the same. In a classroom, it often seems there is more required of a teacher than any single human brain could possibly manage (Feldon, 2007). Consider tutoring a student one on one. This seemingly simple act requires ongoing understanding of the student's cognitive and emotional states, needs, and strengths, and it requires synthesizing all of a teacher's expertise and pedagogical content knowledge to figure out exactly what is optimal at a given time and how best to provide it in the moment (Gess-Newsome, 1999; Koehler & Mishra, 2009).

That is already a difficult task to get right with any consistency. What happens when you add twenty-five more students? In a classroom, not only does every additional student multiply the intrinsic complexity of the task, but new responsibilities emerge: Teachers must monitor classroom engagement, focus distracted students, attend to social and emotional learning, and differentiate instruction for the full range of learners.

Implementing these responsibilities requires extraordinary logistical management. A teacher may be teaching fifth-grade reading, but the reading comprehension skills of students in the class likely span all twelve grade levels (Firmender, Reis, & Sweeny, 2013). And that's just reading comprehension. Students' competencies vary for every subject and skill. A middle school science teacher may invite students to read and discuss a text in groups, but this requires multiple skills: reading, background knowledge in science, and collaboration. Student competencies will range widely for each. Teaching well requires navigating this enormous complexity.

Further, as we've discussed, educators are increasingly loaded with administrative tasks: taking attendance; following up with students who were absent; recording behavioral notes; helping students find the correct file, book, or page number; submitting forms; supervising unstructured time; making sure that students have all the supplies they need and that those supplies are not broken and put away (and, if broken, seeing who broke them and how and deciding how to adjust classroom procedures to reduce the number of materials broken); and on and on. In fact, tasks that may be simple to do in a one-on-one setting—such as finding the relevant document, adjusting the sharing of permissions, and so on—may become difficult when suddenly needed at the same time as everything else.

Educator *and former rocket scientist* Ryan Fuller (2013) may have articulated it perfectly when he says, "Teaching is actually two jobs." One of them, he says, is the teaching, "which would more aptly be called the performance." He continues:

> Every day, a teacher takes the stage to conduct a symphony of human development. A teacher must simultaneously explain the content correctly, make the material interesting, ensure that students are staying on task and understanding the material, and be ready to deal with the curve balls that will be thrown at her every 15 seconds—without flinching—for five hours. (Fuller, 2013)

But he also describes another job:

> [This job] is the one that teachers are familiar with; people who have not taught can pretend it doesn't exist. The tasks involved in this . . . job include lesson planning, grading, calling parents, writing emails, filling out paperwork, going to meetings, attending training, tutoring, and occasionally sponsoring a club or coaching a sport. The time allotted to teachers for this work is usually one hour per workday. But these tasks alone could easily fill a traditional 40-hour work week. (Fuller, 2013)

If this sounds impossible, that's because it is.

Administrative tasks have been shown to crowd out instructional and planning time (Kim, 2019). This is particularly and painfully apparent to teachers. The McKinsey Global Teacher and Student Survey, looking across the United States, the United Kingdom, Canada, and Singapore, finds that teachers work approximately fifty hours a week, with less than half that time spent directly interacting with students (Bryant, Heitz, Sanghvi, & Wagle, 2020; see figure 6.1).

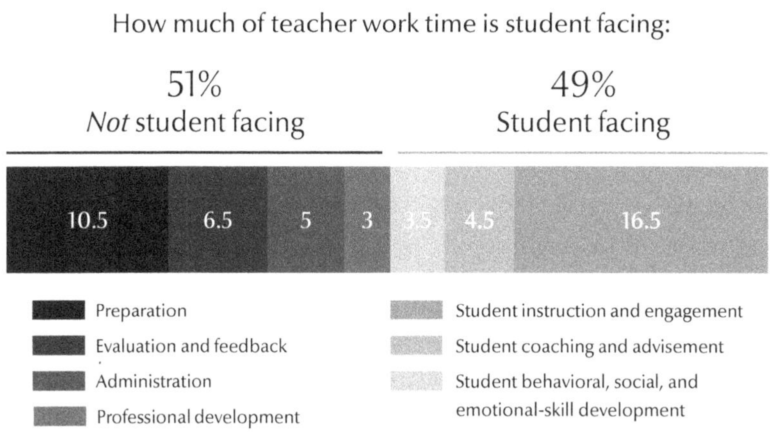

Figure 6.1: *How teachers spend their fifty-hour-average work week.*

The same study finds that not all of this time could be reallocated toward activities that more directly support student learning, but a total of 20–30 percent of it could be. Meanwhile, as confirmation that teachers are still running around chasing administrative needs for the classroom, the teacher resource site Teachers Pay Teachers shows over one hundred results for the search "pencil borrowing," including posters, procedures, labels, sign-out sheets, and more (Teachers Pay Teachers, n.d.).

How do teachers feel about all this? In a 2022 survey of 1,324 teachers, the number-one task that they would choose to spend less time on, if they could, was general administrative work (Merrimack College, 2022; see figure 6.2). As a thirty-year veteran elementary school teacher in Oklahoma puts it:

> My role as a teacher should be to teach the students. I have taught many years and feel each year I get less real time to teach my students and more things piled on my plate to do for documentation or administration duties. I just want to teach my students. (Merrimack College, 2022, p. 17)

In this chapter, we look at how we might leverage technology to not only manage these administrative tasks but also extend our senses and capabilities in the classroom.

Invisible Labor

Some work is obviously work. When we perform a task in exchange for compensation, we have done work and participated in a transactional economy. But what about when work has no compensation or does not take place in a transactional workplace, such as when we care for our children or prepare a meal for our whole family? This work is known as *invisible labor*, a term that researchers developed to describe the work that women traditionally do inside the household that is not compared to the labor men traditionally do outside of the house (Hamblin, Barker, & Arghode, 2020). This invisibility followed from the household into professions traditionally held by women, such as nursing, secretarial work, and flight attending.

In the 2020s, research on invisible labor began to include teaching more often, although it remains sparse (Hamblin et al., 2020; Staudt Willet & He, 2024). This is likely because there is so much *visible* labor

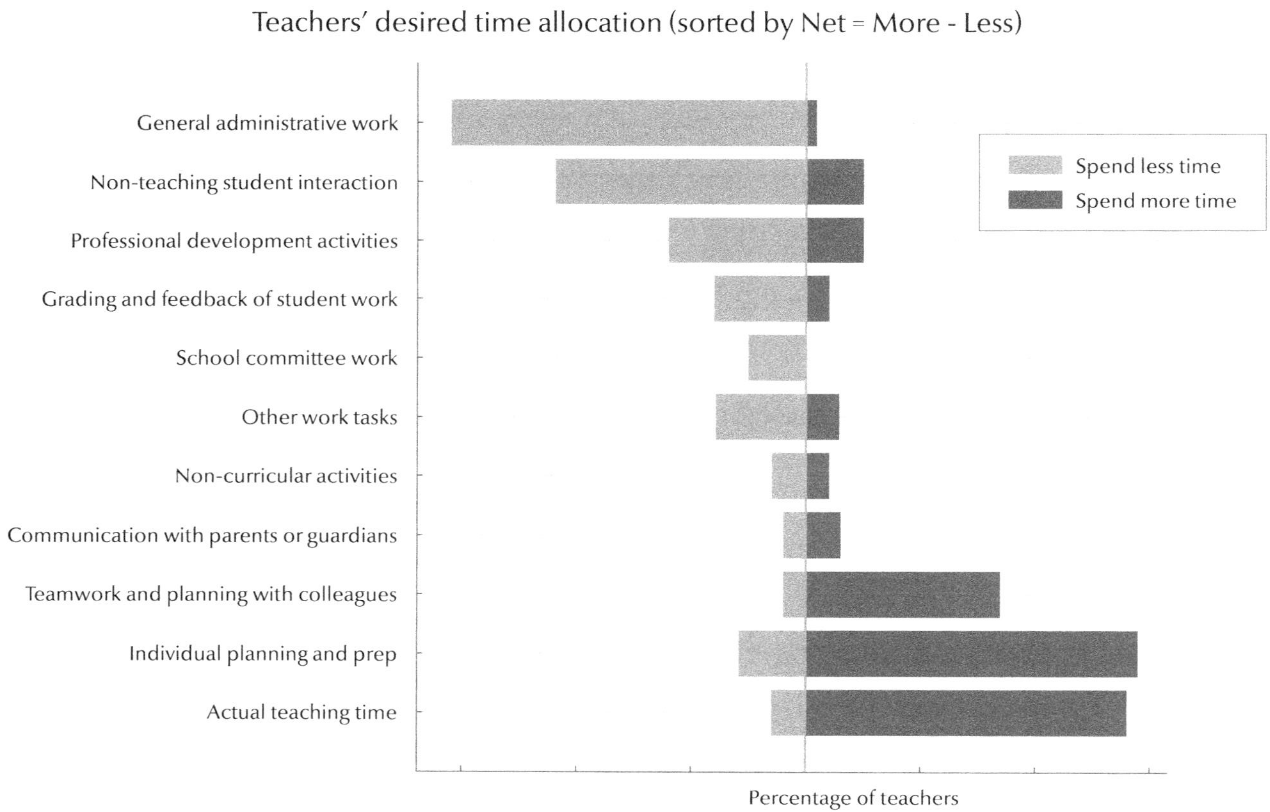

Figure 6.2: *Merrimack College Teacher Survey results of activities teachers would like to spend more and less time doing.*

to teaching that it is hard to conceive that there could additionally be so much *invisible* labor.

So, let's make this work visible. Here is a hypothetical scenario.

Suppose Johnny comes to class late. First, his teacher has to notice that this has happened. Then, she has to record the tardy and follow up with the appropriate people, whether that's the front office staff, the guidance counselor, Johnny's parents, or Johnny himself, each of whom may need a slightly different format of communication. It's difficult for the teacher to stop teaching class to open the appropriate tool, log in with two-factor authentication, and record this tardy, so this means that if Johnny is a student who is always tardy, and his teachers are always too busy to record it, his behavioral pattern will go unrecorded.

Or worse, imagine Johnny says an unkind thing to a classmate, and his teacher overhears while walking around the class and helping students with their work. It takes about a full minute for the teacher to go back to her desk, open the appropriate app, and write a note about Johnny's behavior. If the app is one that interfaces directly with Johnny or with his parents, then the teacher will have to write the note carefully, not just stream of consciousness, which adds another layer of cognitive load. The teacher decides on this occasion to just record the incident on a sticky note and write the longer messages during her lunch break—more work. Every lesson, all it takes is a few such incidents, and vanishingly little time is left for teaching and learning—and this does not include the added cognitive load of switching between tasks (Monsell, 2003).

In both of these examples, the teacher spends time on documentation instead of on teaching. But documentation is not the most important function of a teacher—*teaching* is. It would be a better use of the teacher's time if she could spend class time focused on Johnny's behavior to help him reflect on his actions, self-regulate, and follow up and fix the problem.

This is the situation teachers find themselves in all day, every day. And, at least in our experience, it is not possible to respond consistently well without compromising some other element of work. On a regular basis, teachers must reach out to parents with thoughtful emails, respond to guidance counselors with descriptions of students' strengths and needs, send references that detail students' performance on a variety of tasks, and more. Even when teachers take diligent notes, these notes are not tagged and tabulated, and it can take teachers an entire planning period (or more) to perform these time-sensitive tasks in a timely manner.

Because this thorough process can be so time-consuming, what often happens instead is that when we are asked how our students are doing, we do our best to eyeball an answer. And, because we are human, the feedback ends up being influenced by our human fallibilities. Maybe Johnny happened to have a great day the day the guidance counselor asked. Or maybe he had a particularly bad day the week before. Or maybe Johnny has a sense of humor that the teacher finds distracting, even though it may not be relevant to the need at hand. Or maybe subconscious gender, racial, or cultural differences between Johnny and his teacher affect the teacher's perceptions of Johnny's behavior and intentions (Childs & Wooten, 2023; Dee & Gershenson, 2017).

The cumulative weight of this invisible work taxes teachers' time and cognitive resources. A more reliable method is clearly needed to manage this invisible labor without compromising teachers' primary role in the classroom.

The Ways AI Could Help

Let's imagine an AI company has world-class data privacy and security protocols, so a school allows them to listen to audio data from a classroom. This may be uncomfortable to imagine, but we ask that you suspend your disbelief for a moment. AI hears the teacher say, "Johnny, you're late," and it makes a note. It also records the teacher saying, "Johnny, we've talked about how to act when you feel frustrated with a classmate." AI simply records a summary of these interactions.

Throughout the class, the teacher continues to focus on helping students learn and reflect, and AI records patterns specific to each student as well as trends in the class as a whole. Afterward, AI has composed a draft email both to Johnny's parents and to Johnny, noting what happened, how it was resolved, and how it fits into a broader pattern of Johnny's growth over the course of the school year. It remembers that when this kind of interaction happened two months ago, Johnny was not able to calmly apologize to his group, and it notes the progress in his behavior and self-regulation. For the parents, it might draft a separate email including discussion questions they could use to start a dialogue with Johnny over dinner. For Johnny, a third email might include reflection questions he could answer for homework. Johnny's teacher, at the end of the day, could review these drafts, make changes as needed, and click "send."

Further, AI can keep and scan a cumulative history. When Johnny's guidance counselor asks about "how Johnny's doing" during his teacher's lunch break that day, the teacher can use AI to summarize Johnny's progress throughout his entire school career based on the specific question of the guidance counselor—while she takes a moment to eat her sandwich.

How close are we to this vision? Learning management systems are beginning to integrate AI tools into their product suites, which is a first step toward automating administrative work, but the data privacy concerns of school districts are real. The problem is that, although any single piece of information is harmless, the combination of information specific to a given student could be dangerous in the wrong hands. Therefore, for both technical and ethical reasons, private information about specific underage students being freely recorded in the classroom is unlikely to be mainstream in the near future.

That said, many technology companies have already earned the trust of schools and school districts, and they will likely begin rolling out AI features like this in the coming years. These AI features will be more limited compared to AI-first tools, but they will be a start, and they will soon be integrated into our existing workflows.

The Automation of Grunt Work

Some of the work that an AI administrative assistant would manage is grunt work. But many people use the term *grunt work* in a fast and loose way. For example, in Salman Khan's (2024) book *Brave New Words*, he writes that AI teaching assistants "can do all the grunt work involved in teaching—writing rubrics, giving students feedback on their essays, and drafting student narrative progress reports for parents" (p. 149). Yes, these are tasks that are burdensome because of how much time they take—and we explore here and in part 1 how they can be enhanced by AI—but are they really grunt work when they contribute meaningfully to student learning and relationship building? These are pedagogically important, cognitively rich activities.

Here are some better examples of grunt work: setting up a Google Doc and chasing down each student to make sure they have the correct permissions enabled, copying and pasting twenty-five question stems and answer choices into separate boxes in Kahoot!, and transferring assessment results from one technological tool to another. This is the kind of grunt work teachers are ready to be relieved of.

When ChatGPT came out, Maya was excited to use it to help her with these kinds of tasks. For example, students were requesting more quiz retakes, but they were time intensive to write. So, Maya asked ChatGPT to help her. The questions varied in quality but were good enough as a starting point—until she realized that she would have to painstakingly copy and paste every question stem and answer choice into Canvas by hand. At that point, she decided she would rather do the grunt work of writing the questions herself than the grunt work of copying and pasting for twenty minutes. (This became the motivation for the tool her husband created for her, QuestionWell, which would allow her to export the AI-drafted questions to all her favorite edtech tools.)

Interfacing between tools is invisible labor (Hamblin et al., 2020; Staudt Willet & He, 2024). It is no wonder teachers don't feel valued when tool creators implicitly deem it a good use of teachers' time to copy, paste, and check boxes deep inside of the settings of eighty different profiles across three apps. Rounds of criticism in some educational circles have amplified this message, focusing on the increasing degree to which teachers and students have been reduced, at times, to button pushers, feeding the apps and tools that have become required in their classrooms (Watters, 2021).

In an AI-supported learning environment, AI administrative assistants will manage the real grunt work of teaching, opening teachers up for the more creative aspects of the job. But success depends, at least in part, on how naturally new tools can integrate into teachers' workflows.

A Natural Brain Extension

While we can do simple arithmetic in our heads, if we want to do a more involved mathematics problem, such as multiplying three-digit numbers, we use tools like pencil and paper, or even a calculator. These tools help us externalize our thoughts and make them visible, and by externalizing them, we can more easily manipulate them. One view of these tools and technologies sees them as *extensions* of the mind; in this view, the mind does not stop at the "skin and skull" but rather extends to the outside world of notebooks, manipulatives, computers, and even other people (Clark & Chalmers, 1998).

Consider a simple example of a technology that does not involve the mind, such as a hammer. When we swing a hammer, we are still swinging our own arm. The hammer naturally extends the action we already do; it amplifies our swinging. Its added length and weight increase the force of our arm and extend our strength capabilities many times over. Seen this way, a hammer is a logical extension of our arm.

Today, we are extending not only our bodies but our minds. The following is how Marshall McLuhan (1964) famously conceptualized this idea:

> During the mechanical ages we had extended our bodies in space. Today, after more than a century of electric technology, we have extended our central nervous system itself in a global embrace, abolishing both space and time as far as our planet is concerned. Rapidly, we approach the final phase of the extensions of man—the technological simulation of consciousness,

> when the creative process of knowing will be collectively and corporately extended to the whole of human society. (p. 19)

AI is a self-evident extension of McLuhan's (1964) more inclusive view of the mind, provided that we teachers can figure out an easy and effective way to interface with it.

This is the last mile in the first-mile, last-mile problem from chapter 4. If the first-mile problem is that technology struggles to meet students where they are, the last-mile problem is that technology also struggles to bridge the gap with teachers. For the most part, teachers must use technology on *its* terms, and technology does not extend to and accommodate the specific needs of each teacher.

As a result, teachers' success in working with technology depends on two important considerations: (1) how naturally they experience the technology, and (2) how different technological tools coherently organize together.

First, we experience some technologies as natural or intuitive extensions, whereas others are awkward or forced. Of course, different people may experience technologies differently. Some find voice command technologies a natural extension of speaking, whereas others find it unnatural to stop something they're doing and utter a complete thought to a disembodied computer. But the principle is that some technologies fit within our current work habits, and others require that we change and learn new work habits, some of which feel less natural or efficient. If they are natural, then adoption follows easily. If they are unintuitive, then the technology becomes a labor. The ease of learning the new technologies matters for teachers' busy schedules.

Second, the organization of technologies matters. We all know someone (or maybe we *are* someone) who has four different notebooks in four different places, with different highlights and notes in each, but never quite has the time to review and organize them. How useful are notes that are not accessible? Probably not useful at all. In fact, they may provide a false sense of usefulness (Sparrow, Liu, & Wegner, 2011). Organization and coherence matter because without them, we lose track of our resources and therefore never put them to use. It is the same with our digital technologies. Many tools are accessible to us, but we don't use them due to their lack of operability or interoperability.

The key for a successful brain extension is how naturally and coherently the brain extension transforms the work that needs to be done.

We imagine teachers, for example, using AI voice-activated technology to record what they would have recorded by hand or by going to a particular tab in their browser. But speaking a note and typing a note are more or less equally demanding, at least to those for whom typing is second nature. If they can speak the note while they do something else, that could be time-saving, but is it?

Research in psychology has shown that what we call "multi-tasking" is actually switching between tasks very quickly, and this task switching comes at a large cognitive cost (Monsell, 2003). Speaking, in particular, competes significantly for cognitive resources, especially with tasks involving working memory and attention control (Baddeley, 2003). This leads to a notable reduction in the effectiveness of concurrent tasks (Kunar, Carter, Cohen, & Horowitz, 2008). So, a tool that lets teachers speak instead of write may seem at first to make things easier but instead may just trade one modality for another and provide a false sense of security. This could even backfire as teachers believe that their task should be easier but find instead that it is just as difficult.

AI as Our Administrative Extended Mind

So, how can AI more fundamentally help teachers with administrative tasks?

Sometimes, when we talk about AI and other technologies, we overlook how technology can *transform* the experience of work. The SAMR Model (Romrell, Kidder, & Wood, 2014) aims to represent edtech transformations by grouping technological changes into four categories: (1) substitution, (2) augmentation, (3) modification, or (4) redefinition. These are differences in magnitude: *how* technology changes the work, not binary descriptions of *whether* technology is used. It's the difference between a simple digital to-do list that you can manually add items to—which is basically a *substitution* for a paper to-do list but on a computer—and an AI-generated to-do list that reads your email and extracts and updates to-do items and their status and urgency, effectively *redefining* the work by offloading the executive function to a machine.

This is the kind of augmentation and modification that has the potential to change teachers' lives.

Let's begin with the administrative tasks associated with classroom management. As we previously illustrated, when there is a disruption, not only do teachers need to deal with the disruption, help the student learn in that moment, and get the rest of the class back on track, but they also have to record the incident. If there are more than a handful of incidents per class (and there often are), it becomes necessary to take written notes. This is especially important because 84 percent of public schools reported an increase in student misbehavior since 2020 (National Center for Education Statistics, 2022). Taking better notes of classroom disruptions not only ticks a box but also—and more importantly—provides invaluable information for student support systems and for meetings with parents, support staff, and psychological professionals.

But taking notes on classroom disruptions is just one of many administrative tasks teachers must do in addition to planning and teaching. Estimates place the number of administrative tasks a teacher engages in during the day near twenty-six, and this is in no way comprehensive (Blatchford et al., 2009).

Transforming these tasks from manual labor to automated, redefined experiences can happen in a variety of ways. An adaptation of the SAMR Model, table 6.1 (page 186) shows a handful of examples of administrative tasks and how they could be enhanced and transformed by AI.

We caution that the transformation in this table is not inherently better than enhancement. We provide these examples to expand the imagination of possibilities, not to make a value judgment on a particular approach. Just like it is sometimes preferable for students to hike up the mountain when they are learning in classes (see chapter 5), so, too, is it sometimes preferable for teachers to complete administrative work manually. Each teacher and leader should make their own decisions about which elements of their work are best maintained by hand and which might benefit from a full transformation and utilization of AI.

Table 6.1: *How Administrative Tasks Could Be Enhanced and Transformed by AI*

	Without AI	Enhanced by AI	Transformed by AI
Completing Paperwork	Teachers manually complete forms.	AI assists by prefilling known student information and checking data accuracy.	AI automatically generates paperwork, requiring minimal manual input. Teachers review and approve.
Managing Student Data	Teachers manually enter, update, and retrieve student data from databases.	AI prompts teachers with reminders to update records and highlights unusual data points or trends.	AI continuously updates, analyzes, and presents student data insights to teachers, proactively identifying patterns and suggesting interventions.
Tracking Absences	Teachers manually track attendance, contact parents, and manage follow-up tasks.	AI alerts teachers about attendance irregularities and drafts communication templates for outreach.	AI autonomously tracks attendance, communicates with stakeholders based on school communication practices, and escalates issues needing teacher intervention.
Taking Meeting Minutes	Teachers take manual notes and coordinate follow-ups.	AI generates automated notes or summaries from speech-to-text, flagging important action points.	AI compiles minutes, identifies action items, autonomously coordinates follow-ups with stakeholders, tracks progress of important action points, and produces draft reports for relevant stakeholders. Teachers review and approve at each step.
Tracking Classroom Supplies	Teachers manually monitor supplies, maintaining inventory logs and making replenishment requests.	AI provides digital inventory tracking and sends notifications when supplies run low.	AI predicts future supply needs, automatically managing inventory, budgeting, and ordering supplies, with minimal teacher oversight.
Coordinating Field Trip Logistics	Teachers arrange transportation, permission slips, chaperones, and emergency plans manually.	AI generates checklists, manages RSVPs, sends reminders, and helps schedule transportation and meals.	AI supports the full logistics chain, automatically coordinating with vendors, families, and administrators while monitoring risk and compliance.
Updating Classroom Website or Newsletter	Teachers manually upload content, announcements, and resources.	AI recommends and formats updates based on the curriculum's pacing, students' questions, and upcoming deadlines.	AI continuously updates the class site and weekly email with lesson summaries, reminders, and changes to calendars and timelines.

AI in Practice: Try This

With permission, record meetings you are in and have AI transcribe, summarize, and extract action items from them. Does this make it any easier to pay attention? Does it make it harder? And afterward, do you revisit the notes and action items?

Look at the AI tools that already exist in your student information system or learning management system. These systems already have secure access to student data and will likely be building AI features in the coming years.

Experiment with the AI features built into your email system. If given enough context, can AI draft an email and save you a few minutes? Are there certain email templates you can save as drafts and adjust only slightly each time?

Task and Tool Management

How technological tools are designed to work with other technological tools makes an enormous difference in whether teachers adopt them successfully.

Some solutions focus on adding new tools to a teacher's toolbox. This makes sense, especially when we consider that the mind already depends on outside tools or resources for much of what it does. However, individual tools often simply add to the laundry list of different places to remember to record different things. In other words, without a well-organized toolbox, giving teachers more tools means every tool is harder to find, and the work becomes more chaotic and difficult to accomplish, like a word problem with extraneous information added in to make it more difficult to solve.

So, let's go back to the example of a hammer. If you can't find your hammer because you have too many tools and they are hard to keep track of, you will probably pick a less-ideal tool, even if the hammer is perfectly designed to be an extension of your arm. If most of your tasks involve hammering, then having all the other tools simply clutters your toolbox and makes it harder to do your job. LearnPlatform (n.d.) by Instructure (the company behind Canvas) reports that the average number of edtech tools accessed per school district over the duration of the 2023–2024 school year was 2,739. They ask the question explicitly: "As the average number of tools districts access steadily rises, how will organizations *effectively maintain the increasingly diverse ecosystem of tools* required to reach teachers and learners?" (emphasis in the original; LearnPlatform, n.d.). They leave the question unanswered.

One failure mode is having too many and too disorganized tools; the opposite failure mode is having too few. If all you had was a hammer, you would have a hard time with tasks that required screwdrivers and drills. But which is the lesser of the two imperfect scenarios? Is simpler better? Is having a tool that is good enough most of the time preferable to chaotically switching tools all the time? Or is it better to have every possible tool available but not be excellent at any of them and often struggle to integrate them with each other?

This conflict is why teachers sometimes seem so reluctant to learn new tools. Even if we go to a professional development session for every new piece of technology and have a beginner's understanding of how to use each one, there are simply too many tools—most of which are complicated and not interoperable, many of which come and go within a few years, and all of which require experience to master. A teacher is usually focused on mastering classroom management, pedagogy, student names and needs, and other more urgent priorities. Mastering a wide range of technological tools on top of those concerns requires years of experience. Further (and unfortunately), the tools change more quickly than the teachers can learn them. In our experience, this is why teachers typically pick whichever tools seem most useful, get good at using those, and, if only for the sake of time, ignore the rest.

If we think of technology as extending the mind, the problem becomes clear. Adding more parts to the mind may add processing power, but it also creates the challenge of coordinating and synthesizing new information, which is arguably a more difficult task. Teachers have been labeled as *Luddites*, *risk averse*, and *resistant to change*, when it is a lack of a coherent, shared vision for technology use that causes new tools to add to teacher workload rather than decrease it (Howard & Mozejko, 2015).

Review and Next Steps

Teaching is harder than rocket science—not because teachers aren't capable but because they're asked to do too much. The core work of instruction—designing and facilitating learning, responding to students' needs, and managing the dynamics of a classroom—is cognitively demanding on its own. Add in the endless administrative load—taking attendance, writing notes on students' behavior, troubleshooting technology, emailing families, and so on—and the job becomes not just exhausting but unsustainable. Teachers are expected to be performers, managers, caregivers, and clerks, often switching roles every few seconds.

What makes this workload especially difficult is how much of it is invisible. Teachers don't just teach—they chase down login issues, write careful notes about misbehavior, make sure the right students are in the right Google Docs at the right time. This hidden layer of work fragments attention and erodes capacity. Even high-value tasks like giving feedback or writing progress reports become draining when multiplied across dozens of students and scattered across disconnected platforms. It's not the cognitive richness that grinds teachers down—it's the logistical tedium and lack of support for execution.

AI offers the possibility of relief. Imagine AI that notices when a student is late, drafts a message to their parents, logs the incident, and links it to a broader pattern—all while the teacher stays focused on the lesson. Or imagine one that compiles student data and autogenerates summaries for counselors or Individualized Education Program (IEP) meetings. When AI transforms administrative work from scratch built to review-and-send, it does more than save time—it protects the teacher's limited cognitive bandwidth. It does more than make teachers more efficient; it restores purpose.

But for AI to help rather than hinder, it must integrate smoothly into teachers' workflows. Most technology today fails because it adds friction. Teachers don't need more tools—they need fewer, better-organized, more interoperable ones. When AI is designed as a natural extension of how teachers already think and work, it can help offload invisible labor and allow teachers to focus on what matters most: their students. The goal is to make more room for the deeply human work of teaching. And this is what we explore further in chapter 7.

This section details some next steps to consider for educators, school leaders, and those working in the technology industry.

For Educators

The following are some considerations for educators related to AI for administrative classroom work.

Conduct an administrative audit. Notice how long you spend on administrative tasks per day. Consider tallying these tasks at the end of each class or day for a few days. Are there any processes you can streamline? Would the workflow benefit from organization, such as having separate browser profiles for online tasks, each with their own bookmarks? How many are not technology intermediated? Consider approaching your building's instructional technologist with your findings and questions about how to streamline the work.

Stay up to date on AI features of existing tools. Work with your instructional technologist to receive regular updates. As these features are rolled out, don't be afraid to be the first to try them, see if they work, and report back. The way they will improve is with your practical feedback!

Advocate for tools you love. Overall, administrators try to only spend money on the things teachers will actually use. That said, they don't always have the best ways to keep track of what those tools are. Be vocal if you like a certain tool. Push your administration to subsidize it and train others.

For School Leaders

The following are some considerations for school leaders related to AI for classroom administrative work.

Design the toolbox—don't just add tools. Rather than considering the possible effects of any individual new tool, consider how you are building the toolbox for the teachers in your schools. How many platforms are they switching between? Is there any way to smooth out or even automate those connection points? Ask them specific questions about how long different tasks take and dig into why. For example, how long does writing a progress report comment take? Multiply that by the number of students. What tools do teachers use to gather all the information they need to even start writing that comment? The whole workflow is often more burdensome than the sum of its parts, and that extra work is invisible labor that can lead to burnout (Hamblin et al., 2020; Staudt Willet & He, 2024).

Consider each of the following use cases and look for tools to help your teachers automate the grunt work. There are many different applications of AI for administrative work. The following list is meant as a starting point. For each use case, consider how it would fit into teachers' workflows. Would they have to manually copy and paste what AI writes? How long will this ultimately take? Look up YouTube tutorials if you're not sure. Make sure you are not replacing tedious work with grunt work.

- **Accessibility**
  - Making materials accessible for English learners
    - Translating to another language
    - Changing reading level
    - Choosing vocabulary
  - Generating notes for students with dysgraphia
  - Creating IEP documentation
- **Technology Management**
  - Opening and closing apps and navigating to the right files and tabs
  - Managing permissions on Google Docs, making sure the Canvas assignments have a way to submit, and so on

- **Communication**
    - Communicating with parents, learning specialists, and administrators
    - Writing newsletters and progress reports
    - Catching up students who missed class
- **Behavioral Patterns Over Time**
    - Tracking tardiness and absences
    - Noting students who are unprepared (such as needing to borrow a charger or pencil)
    - Tracking bathroom sign-outs

For Technologists

The following are some considerations for those working in the technology industry related to AI for classroom administrative work.

Dream big but stay grounded. It is going to take vision and innovation to get from here to a world where AI has truly transformed the teaching experience for the better. Not being stuck in the trenches, so to speak, can be liberating and allow you to think creatively. But be cautious that your thinking outside the box doesn't lead to inventions that won't then fit into the box. In other words, make sure you are constantly talking to teachers (and paying them for their time). This will keep you grounded in the customer's experience. Listen to their pain points and ground your innovations in teachers' experiences.

Make your tool play nice with other tools. One way to compete against other technological platforms is to make it so that your users must transition entirely to your platform. This assumes a zero-sum world; if a teacher is on one platform, then they are not on another. As a result, companies try to make one-stop solutions that do everything but end up providing all the services at a medium level. But teachers don't want a zero-sum model; they want highly effective, easy-to-use tools for learning. If tools work well together, teachers are happy to use many different ones, especially if each one is good at its one specific function and interfaces nicely with the other tools.

Think about teachers' experiences outside of your one tool. Talk with teachers about what other tools they're using—not to get insight into competitors but to understand how your tool fits into their workflows. Look for ways to smooth out the transition points in their workflows and make sure you're not de facto insulting teachers by leaving it up to them to do the less creative, more boring work, like copying and reformatting, because they are locked into your tool.

What if technology companies could divide up the work, each focus on one useful function, and then intentionally design for teachers to easily mix and match their favorite tools? This would create a better experience for everyone.

CHAPTER 7

The Teaching Assistant

Ongoing, Embedded, Collaborative Learning for Teachers

The assumption that teachers can create and maintain those conditions which make school learning . . . stimulating for children, without those same conditions existing for teachers, has no warrant in the history of man.

—SEYMOUR SARASON

In the famous play *Cyrano de Bergerac* (Rostand, 1898), the handsome but hapless Christian de Neuvillette falls hopelessly in love with the brilliant and beautiful Roxane. Roxane is enamored by Christian's appearance, but unfortunately, Christian is clumsy with words. So, to woo her, he receives help from Roxane's distant cousin, the eloquent title character, Cyrano.

Cyrano writes poetic love letters for Christian, and Roxane grows fond of Christian through his letters. Encouraged, Christian then tries to court Roxane with his own words. He goes to her home and calls her to her balcony, but when he speaks without help, he embarrasses himself and angers Roxane. So, Cyrano speaks for him again, this time aloud, hiding in the bushes nearby, moving Roxane to love and tears with his inspired poetry and then coaching Christian to climb to Roxane for a kiss.

We hope we are not as hapless in teaching as Christian de Neuvillette is in love, but we also would be glad to have an always-on-call Cyrano de Bergerac, a teaching assistant to help us understand our classrooms and provide the right words or references at the right time.

The purpose of this chapter is to explore this question: How might we draw support from an AI teaching assistant that works like Cyrano, helping us not only with managing the invisible labor of a classroom but also with the human performance part of teaching, too—engaging with a room full of students in service of our learning objectives?

Whether teaching is more or less complicated than courtly love in 17th century France, we'll leave for others to decide.

As a real-time support, Cyrano de Bergerac stands in sharp contrast to Jafar in *Aladdin* or the wicked adviser Wormtongue in *Lord of the Rings*. Both of those figures whisper into the ears of their rulers, corrupting their leadership and pushing their kingdoms to near ruin. In the framework from The Rithm Project shown in chapter 4 (figure 4.2, page 108), Cyrano would lean toward being characterized as prosocial (since he strengthens Christian's engagement with the world, though he creates some dependency), and Wormtongue and Jafar would be antisocial (since they lead their kings to retreat further from the world).

What would a prosocial teaching assistant look like in the classroom? How might a Cyrano support us as teachers *during class* in a way that *strengthens* our relationships with our students? And how might a Cyrano coach us *after class* and help us reflect on what worked and what we might want to try in our next class? These are the questions for this chapter.

Real-Time Teaching Assistants (During Class)

You can be my wingman anytime.

—TOP GUN

Is it even plausible that a teacher might introduce an AI device to the classroom and not be distracted by it? This might seem unlikely, given the way phone notifications fragment our attention. However, precedents for real-time human partnerships in the classroom can help guide our thinking in three ways: (1) *Teaching assistants* have long been a model in higher education for classroom support, (2) *team teaching* has served primary and secondary educators for decades, and (3) *instructional coaches* have grown in popularity in the 21st century.

These roles and relationships offer an additional set of eyes and ears in the classroom, sometimes to support the lead teacher and other times to support students. Rather than making the classroom more fragmented, these models of human collaboration allow a lead teacher to either learn from an additional presence in the room in real time or delegate parts of the learning experience, reducing rather than adding to their cognitive load.

In chapter 6, we explored the ways AI assistants remove some of the logistic burdens of the classroom. Now, we look at the ways AI teaching assistants might support the pedagogical and relational experience of the classroom.

AI Assistant as Teaching Assistant or Team Teacher

Are two teachers better than one? At least in the context of team teaching, it appears so. A balm for the isolation of the classroom, team teaching provides teachers with professional partnership and social support. When given support from administrators, trained on effective methods for team teaching, and provided adequate time to plan, teachers report in surveys that they value the professional growth, shared responsibility, and reduced isolation of team teaching. However, the reverse is also true: When *not* given adequate support, training, or time to plan, teachers prefer to teach alone (Cook & Friend, 1995). But when the conditions are right, teachers show a strengthened professional identity and even describe team teaching as the best professional growth experience of their careers (De Backer, Schelfhout, Simons, Vandervieren, & Rivera Espejo, 2023; Walther-Thomas, 1997).

What has team teaching historically looked like? The following is a brief synopsis of different models (Cook & Friend, 1995).

- **One teacher, one assistant:** One teacher leads instruction while the other provides support by circulating the room, assisting students, and managing classroom needs.

- **Station teaching:** Teachers divide students into small groups and rotate them through different learning stations, with each teacher responsible for the instruction at a specific station.
- **Parallel teaching:** The class is split into two smaller groups, and each teacher simultaneously delivers the same lesson to their respective group to improve engagement and differentiation.
- **Alternative teaching:** One teacher leads instruction for most of the class while the other provides targeted instruction for a small group of students for remediation, enrichment, or special support.
- **Team teaching:** Both teachers actively co-instruct the entire class at the same time, sharing responsibility for lesson delivery, discussion, and student engagement.

The most common team teaching model has been the one teacher, one assistant model (Scruggs, Mastropieri, & McDuffie, 2007). What might this model look like now, when every student has access to an AI version of the circulating teaching assistant? Maybe it would seem like a distraction in the hands of each student, but it is not dissimilar to each student having a textbook on their desk. When students have questions, they may look to the book for support, and when the teacher requires the attention of the entire class, all books close. Similarly, saying "laptops down" draws students' attention from their devices back to the teacher. So it could be with an AI teaching assistant. When confused during independent work time in class, students might ask for assistance from an AI teaching assistant and return to the class when ready or called back. With appropriate classroom management, this could provide support for students and teachers alike.

This form of team teaching with an AI teaching assistant—one in which the teacher leads the lesson and students seek and gather extra support from the AI teaching assistant when needed—is not a large leap from current classroom paradigms. The pedagogy remains mostly the same, but the tools shift. A more significant departure, however, is the AI as an instructional coach.

AI Assistant as Instructional Coach

Traditionally, instructional coaching involves an observer taking notes in a classroom and then conferring with the teacher after class to debrief and set goals for the next class. With wireless technologies for communication, however, and with cameras and microphones as additional sets of unobtrusive eyes and ears, researchers have been exploring new and different models, including teachers wearing earpieces through which coaches can speak with them during a class. Much closer to the model of Cyrano de Bergerac, variations on this real-time, bug-in-ear (BIE) approach to coaching are poised to grow in an age of AI.

What has research into this kind of real-time support yielded? After a period of acclimation, teachers with BIE technology feel more confident and skilled, enjoying the ability to alter and improve their practice immediately based on real-time feedback (Rock et al., 2009; Rock et al., 2014; Scheeler, McKinnon, & Stout, 2012). While initially nervous about the experience, teachers ultimately described it as empowering due to the positive suggestions and affirmations they gathered from the support of a real-time coach. And, most importantly, BIE feedback during teaching yields more effective practices, not only during the course of the BIE implementation but afterward, too.

This kind of real-time support has not yet been studied extensively with AI on the other end of the BIE channel, but early research has emerged on live AI support in the context of virtual tutors. In a Stanford University study, for example, 1,800 students were assigned to nine hundred tutors, some of whom were randomly selected to have access to AI tutor support (Maslej et al., 2024). In this virtual context, tutors could push a button and receive multiple recommendations for how they might respond to students' questions. These live AI suggestions for the tutor included a proposed question, hint, or explanation of a concept, including possible scripts for the question, hint, or explanation. The tutor could then choose to deliver one of the recommended approaches or something else entirely. Students who were assigned to tutors with access to AI support were more likely to master topics than those who were not, as evidenced by performance on exit tickets from the tutoring sessions (Wang, Ribeiro, Robinson, Loeb, & Demszky, 2024). Notably, exit ticket performance improved the most for students of weak tutors who used AI and less significantly for students of stronger tutors who used AI.

So, AI has been used effectively to coach tutors during live tutoring sessions, and teachers have found live feedback while teaching to measurably improve not only their practice but also their feelings of confidence and effectiveness. What might it look like to bring these developments together in an AI-powered, real-time support for teachers in the classroom, and what practical needs would this meet?

Classroom Tasks (During Class)

So, AI has the potential to help teachers with live pedagogical support in the classroom. But what, specifically, might that help focus on? Teachers and students manage numerous tasks during any class period, and the following sections break down how AI might support such tasks.

Practical Help for Teachers

Over the duration of a class, teachers juggle a series of priorities.

- Enacting the class plan
- Monitoring student engagement in the learning experience
- Adjusting pedagogy based on student engagement
- Assessing student progress toward objectives
- Adjusting plans based on student engagement and progress
- Concluding class
- Setting up homework and the next class

AI might support teachers across these priorities. Table 7.1 shows a selection of areas where teachers engage in pedagogical practices that might be aided by an AI teaching assistant.

Each of these actions provides valuable information for the teacher in support of their effective pedagogy. In the spirit of prosocial AI development, AI teaching assistants performing these functions serve not to replace teachers in the classroom but to support them by

Table 7.1: Specific Teacher Priorities During a Class and AI Support Opportunities

Teacher Priority	AI Teaching Assistant Support Opportunities
Engage students and monitor participation.	• Monitor student participation through voice recognition or visual tracking. • Send notifications to the teacher to engage a particular student at a particular time, as needed.
Track wait time after asking a question.	• Observe when a teacher is asking a question. • Measure pauses after the teacher asks a question. • When appropriate, remind the teacher of wait time to allow students time to think.
Foster inquiry.	• Observe when a teacher is asking a question. • Identify whether the question a teacher asks is open-ended or closed. • Provide revised versions of the question.
Assess progress and help students achieve desired learning outcomes.	• Transcribe and analyze the dialogue in a classroom. • Assess collective or individual progress toward identified learning goals. • Draw a teacher's attention to students in need of support. • Provide possible modifications to the class plan.
Adjust plan when student needs aren't being met.	• Identify common misconceptions stated by students during class. • Propose alternate explanations, activities, models, or engagement mechanisms.
Identify when the lesson is off track or falling behind.	• Notify the teacher about the time remaining in the class period. • Propose options to adjust the lesson, homework, or the next class.
Conclude class effectively and on time.	• Summarize class progress in several sentences. • Offer options for revised exit tickets. • Adjust homework plan based on class progress.

strengthening the relationships at the heart of the instructional core—those among the teacher, students, and curriculum.

Practical Help for Students

AI teaching assistants in the classroom also support students. Just as a student might look up notes they had previously taken, refer to a passage in a textbook, or check a piece of information online, so might they interact with an AI teaching assistant. The AI teaching assistant might answer a question, guide the student through part of a lesson that the student doesn't comprehend, or even support a student in building the confidence to participate in a class discussion. What might these supports look like? Table 7.2 shows just a few ways in which an AI teaching assistant might support students in class.

Table 7.2: *Specific Student Needs During a Class and AI Support Opportunities*

Student Need	AI Teaching Assistant Support Opportunities
Guidance on work produced in class	• Provide first-pass feedback based on specific criteria or the student's question.
Class content comprehension	• Provide information, offer context, and relate to previous lessons or understanding. • Offer examples from different contexts. • Generate analogies connecting class content to the student's interests or expertise. • Create practice questions, problems, scenarios, or exercises to deepen or expand mastery.
Logistical support	• Provide technological guidance for students similar to the support for teachers mentioned in chapter 6.
Participation in a discussion	• Draw on the student's previous work to prompt them to realize they have a question or are interested in a line of inquiry that they might bring up to the class.

In each of these cases, the AI teaching assistant offers not only another set of eyes and ears but also a pedagogical partnership with the lead teacher. The relationships among the lead teacher, the students, and the curriculum remain not only intact but stronger for having an additional facilitator. If, for example, a student in a mathematics class is struggling with a particular concept, the student might ask the AI teaching assistant to offer additional explanations of the concept. After the exchange, the student may have stronger mastery, and the teacher can know from the AI teaching assistant what area was challenging, what helped, and what level of competence and confidence the student has now, thereby informing their next exchange.

This exchange, it is worth noting, is one in which a student is working *one-on-one* with an AI teaching assistant. As more individual content and skill mastery move outside of the classroom, more one-on-one exchanges will happen outside the classroom, and more student engagement inside the classroom will involve collaborative meaning making and shared experiences between students and the teacher. Success at supporting these kinds of collaborative social experiences in the classroom will be even more challenging than success at supporting students individually. Can an AI teaching assistant support a small breakout group of students in a class? Probably not in any consistent way, given the complex social dynamics of children and adolescents. But beginning to attempt it will require at least the areas of development noted back in chapter 4. Those are explorations for another day.

So, robust opportunities exist for AI teaching assistants to support both teachers and students *during* class. *After* class, opportunities for teacher reflection and growth are near limitless.

Feedback and Reflection (After Class)

Feedback for teachers during class may be limited by cognitive load, but feedback after class is limited only by the questions they have the imagination to ask. With transcripts and recordings of teaching more readily available than ever before, and with an AI teaching assistant ready to help parse them, teachers need only to choose what they would like to work on.

The literature on effective feedback from observations and instructional coaches is vast, so here, we will focus specifically on two elements of after-class feedback that we believe are most important for building cultures of ongoing feedback: (1) goal setting and (2) structural support.

Teacher-Driven Goal Setting

Feedback for teachers is most effective when it supports goals they have set for themselves (Criss, Konrad, Alber-Morgan, Brock, & Harris, 2024). Further, teachers who are given autonomy to define both their goals and the processes to pursue them report feeling respected as professionals and express greater satisfaction with their professional development (Walker, 2022). In the first large-scale quantitative study of teacher autonomy in England, it was the teachers who could set their own professional development goals who reported the highest level of job satisfaction and greater levels of intention to remain in teaching (Worth & Van den Brande, 2020). Autonomy like this fosters intrinsic motivation, strengthens teachers' sense of agency, and sustains their enthusiasm for continued growth.

Helpfully, teachers have no shortage of questions about their work. Given the opportunity to set their own goals for debriefing sessions with an AI teaching assistant after classes, teachers might choose to inquire about a host of topics, including—but certainly not limited to—the following.

- "Did all students participate? How much did they participate?"
- "In what ways did they participate? With questions? Answers? Distracting comments?"
- "What kinds of questions did I ask? Open-ended or closed?"
- "Did I pause for sufficient wait time after I asked questions?"
- "Were my transitions between activities smooth and efficient? How could I optimize them further?"
- "How effectively did I check for student understanding during the lesson?"
- "How did I handle student misconceptions or incorrect answers? What are some remaining misconceptions my students might have?"
- "Did I foster a supportive classroom atmosphere? What specific phrases did I say that contributed to or detracted from this atmosphere?"
- "Was my feedback for students relevant, timely, clear, and actionable?"
- "Were there moments I could have adapted my lesson to better meet student needs?"

- "Did I effectively use board space or other visual media?"
- "Did I provide enough time for the activity?"
- "How effective was the end-of-class sequence?"
- "Did I achieve my goals for the class?"

Teachers setting their own goals is foundational for sustained growth, but feedback becomes even more powerful when it aligns with broader aims, whether at the departmental, institutional, or student level (Park, Takahashi, & White, 2014). An AI teaching assistant can help broker this alignment. For example, if a school is prioritizing equitable participation, teachers might choose to set personal goals on measuring and improving student voice in their classrooms. Because AI can process transcripts at scale, it can provide individual teachers with feedback that not only serves their personal growth but also draws on priorities set by a schoolwide effort. This creates a productive balance: Teachers retain ownership of their goals, and their inquiries are situated within shared priorities, strengthening both individual practice and organizational culture.

AI-based feedback can be even more effective when it is complemented by other forms of feedback. Teacher-driven exchanges with AI can be paired with, for example, student surveys, peer observations, or observations aligned with departmental benchmarks. Together, these forms of feedback can create a rich portrait of how the classroom works. In practice, this might look like a teacher surveying students about their perception of class discussions and creating the hypothesis that they could improve the quality of discussions if they leaned more toward open-ended questions than closed, then asking AI to track the ratio of open-ended to closed questions in a lesson to become more aware of this feature of their class. By substantiating self-chosen inquiries with different sources of feedback, teachers can ensure their growth remains self-directed while still informed by the perspectives of those most affected by their teaching.

AI in Practice: Try This

What does AI-provided feedback look like in practice? For any of the previously listed questions or any of the actions we take in our classes every day, we might direct our AI teaching assistant to gather notes, share feedback, and prompt reflection on each one with the following actions. Try working with a recording of class to explore any of the following pedagogical objectives. Some will require more advanced tools, but others might be accomplished with standard AI tools. Be sure to be attentive to personally identifying information for students; remove it when necessary or appropriate. Following is a list of possible directions to provide to AI organized around the teacher priorities from earlier in this chapter (page 201).

1. **Enact plan.**
    - *Summarize class*—Provide an overview of what transpired in class, highlighting variations and changes from the original plan.
    - *Extract key discussion points*—Identify what was most frequently or heavily discussed and summarize major student contributions.
2. **Monitor student engagement.**
    - *Track student participation*—Observe and report each student's participation in relation to objectives and competencies, breaking the participation down by degree and type.
    - *Identify engagement trends*—Analyze engagement levels across different parts of the lesson, and identify which teaching methods elicit the most prevalent and high-quality student participation.
3. **Adjust pedagogy.**
    - *Classify questions*—Transcribe each question a teacher asks, and classify the type of question, reporting back to the teacher the frequency of each question type.
    - *Assess wait time*—Measure the average wait time after each question in class is asked.

- *Analyze student-teacher interactions*—Measure the proportion of teacher versus student speaking time, and identify patterns in questioning and student responses.

4. **Assess student progress.**
   - *Monitor comprehension*—Identify patterns of misunderstanding or confusion.
   - *Assess mastery*—Observe and record student participation specifically related to objectives and competencies.
   - *Recommend groupings*—Suggest groupings for future activities based on student strengths, weaknesses, or engagement patterns.
5. **Adjust plan.**
   - *Track differences between sections*—Record differences in progress across different courses in terms of topics covered, mastery demonstrated, and misconceptions predicted. Then, propose differentiations for each class to bring the sections up to speed in the next class.
   - *Suggest pacing adjustments*—Analyze lesson pacing to date to suggest sections of future plans that may need more or less time.
6. **Conclude.**
   - *Summarize student takeaways*—Provide insights on what students likely retained.

Teacher-driven goal setting and reflection, facilitated by AI teaching assistants, make after-class feedback personal, flexible, and actionable. By choosing their own questions and pairing them with institutional priorities, teachers gather feedback that is relevant to their growth while contributing to collective improvement. This dual focus strengthens not only individual practice but also the culture of shared learning within schools.

Still, is there time and support for this? Successful AI-assisted teacher-driven feedback depends on structures that make this work sustainable. Teachers need time to review feedback and a culture that values experimentation and reflection. This is where administrators play a critical role in creating the conditions that allow teachers to pursue their goals effectively.

Structural Support for Teacher Growth

Technology entrepreneurs often say that vitamins don't sell, but painkillers do. If a product solves a real pain point—a "hair on fire" problem—schools will pay for it. If it's merely a nice-to-have, then adoption and sustainability are far less likely. Professional development tools, unfortunately, often are treated like vitamins: something theoretically good for teachers but crowded out by the urgency of more immediate needs.

Professional development makes the most difference when it is collaborative, ongoing, and practice focused (Darling-Hammond et al., 2009; Desimone, 2011). This only happens when it is job embedded, with administrators prioritizing it. Framing professional growth as *essential* to collective teaching quality—and therefore as a painkiller rather than a vitamin—changes how schools invest in it.

An AI teaching assistant isn't effective if a teacher doesn't have the time or opportunity to use it. For any growth-oriented tools to work, they must be integrated into a continuous professional development program. This requires adequate, dedicated time and a growth-oriented culture.

Time

Integration begins with time. A continuous professional development program requires room in teachers' schedules for collaboration, reflection, and taking action. When professional growth is planned into the school day, teachers are more likely to engage deeply and sustain changes in their practice.

The *Time for Teachers* report finds that students of teachers in schools that offer thirty to one hundred hours per year (or one to three hours per week) of job-embedded teacher professional development during the school day score an average of 21 percentage points higher than peers whose teachers have less time for professional growth (Kaplan, Chan, Farbman, & Novoryta, 2015). Yet, in the United States, fewer than one-quarter of teachers report receiving this minimum threshold. Meaningful professional development cannot be an afterthought—it must be intentionally scheduled and protected with extended blocks of lesson planning, peer observation, and data analysis and a school's treatment of this time as central to its teachers' instructional quality.

The same requirements apply for AI-assisted feedback to take root. AI may make gathering feedback easier, but teachers need regular, dedicated opportunities during the workday to review transcripts, examine AI-generated insights, and translate those reflections into instructional choices. Protected time communicates that growth is not a side project but part of the work. This could mean embedding reflection sessions into the weekly schedule, aligning AI debriefs with professional learning community (PLC) meetings, or carving out short but regular windows during the school day for teachers to engage with each other or with their AI teaching assistants. Without time, even the most sophisticated systems risk becoming nice-to-haves that never get used. With time, they become a foundational component of continuous improvement.

Culture

Teaching is a demanding craft defined by successes and setbacks that unfold every day, and no teacher delivers perfect lessons with perfect consistency. Instead, the complexity of the role often leaves educators stretched in ever-surprising ways (Hlado & Harvankova, 2024; Hlado, Juhaňák, & Harvankova, 2025). It takes humility to face these surprises and setbacks, recognize that growth is ongoing, and seek to continually improve. It would be much easier to be satisfied with something that works most of the time, but great teaching means creating ongoing learning for teachers as well as students, and doing that requires conditions where it is safe to be vulnerable about what is working and what isn't.

Administrators play a central role in cultivating that safety. In schools where psychological safety is nurtured, teachers are more likely to take risks, open their practice to others, and embrace feedback as an opportunity for growth (Shahid & Din, 2021). This aligns with research on student learning, as well, as we saw in chapter 3: Feedback that undermines competence or autonomy can backfire, lowering motivation (Kluger & DeNisi, 1996), while feedback that builds competence and affirms autonomy can fuel intrinsic motivation (Deci & Ryan, 1985; Vallerand & Reid, 1984). Administrators who normalize openness, model humility, and frame professional development as collective work establish the trust required for meaningful improvement.

Within this culture, feedback works best when it leads with strengths and offers high-quality information about how we work and how to improve. Affirming progress sustains motivation, while error-focused comments alone can be discouraging. John Hattie and Helen Timperley (2007) found that the most effective feedback helps learners see where they are, how they got there, and what to do next—what we called *high-information feedback* in chapter 3. Administrators who support systems that highlight teachers' strengths and provide clear, actionable insights help develop the kind of culture of growth that fosters teacher and student success.

AI can help when embedded in this kind of culture, but it must be designed for that purpose—affirming strengths, offering precise and actionable insights, and respecting teachers' autonomy. Otherwise, the constant highlighting of shortcomings risks adding complication to an already complicated workplace. As Seymour B. Sarason (1974) reminds us, teachers cannot build those conditions for learners if they do not also exist for the teachers themselves.

Review and Next Steps

Is the vision of an AI teaching assistant a useful one? Does it solve an actual problem, or is it a technophile's solution in search of a use case? As chapter 1 demonstrated, teachers continue to say *yes* when asked whether they want support that gives them more time to plan, reflect, and grow (Merrimack College, 2022). But, as chapter 6 reminded us, the barrier is not desire but workload. Taking steps to provide time and create the right culture can help meet the needs and desires of teachers.

This chapter explored how AI might serve as a prosocial teaching assistant—sometimes a Cyrano whispering in our ear in real time and other times a coach helping us review after class. We looked at how real-time supports can reduce cognitive load, how after-class feedback can strengthen reflection, and how self-directed goals, paired with clear tactics, can make teacher growth both personal and actionable. We also saw that for these supports to work, they must be framed not as add-ons but as parts of a continuous professional development culture; they must be embedded in schedules, aligned with school priorities, and nurtured by a culture of collaboration.

For our part as teachers, we have opportunities to set goals and experiment. And for administrators, the opportunity is to create the time and structure for teachers to engage and grow.

Teaching is imperfectible. But the promise in this work is to create the conditions where sustained learning exists for all members of our school communities. This would let us affirm our strengths, identify opportunities for improvement, and participate in ongoing feedback cycles that reinforce both teacher efficacy and agency.

This section details some next steps to consider for educators, school leaders, and those working in the technology industry.

For Educators

The following are some considerations for educators related to AI for teaching assistance.

Start with one class transcript. Start with small experiments. Audio-record one of your classes, generate a transcript, and anonymize the student names. Then, use AI to examine aspects of your teaching that you are most curious about—for instance, the number of open-ended versus closed questions you asked, the average wait time after posing a question, the distribution of students' participation, or the key misconceptions students revealed. Once you have the feedback, choose one insight to act on in your next class and repeat the cycle with another recording. If you wish, consider inviting a colleague to look at the same transcript or reflect on AI's findings with you to layer human and AI perspectives, deepening your understanding.

Pair AI feedback with student or colleague feedback. After using AI to analyze a transcript, compare its insights with student surveys or exit tickets. For example, if AI notes that most of your questions were closed, ask students how engaged they felt in the discussion. Consider focusing on asking more open-ended questions in your next class and surveying the students again. These kinds of cross-checks build a richer picture of the classroom and keep students' voices at the center.

For School Leaders

The following are some considerations for school leaders related to AI for teaching assistance.

Identify areas for teacher coordination around growth. What are your school's teaching and learning priorities, and where might AI feedback for teachers fit in? Invite faculty to develop a list of teaching and learning areas they might wish to explore, then set aside time in your professional development calendar for teachers to convene, identify how they might gather data on the topic, and develop approaches whereby AI might help them analyze and reflect on that data. Involving teachers in the process of setting priorities and goals will help them feel ownership of the process and increase their motivation to experiment.

Start with person-to-person collaboration. As part of an ongoing professional development program, explore the viability of testing out team teaching cycles, perhaps only for a unit at a time. Alternatively, consider other observational cycles in which teachers observe a colleague at least one class per month. For either of these (or others), consider experimenting with protocols focused on specific classroom needs, especially those that produce artifacts for professional reflection. Live feedback and observations for postclass reflection don't need to begin with technology. What areas of observation and review do teachers find useful? Try asking, then consider which might be supported with technology. Is it identifying patterns of student misconceptions, balancing teacher talk versus student talk, analyzing question types, or something else entirely?

Identify a small group of early adopters or experimenters. Organizations that evolve often have a team whose purpose is to actively, continuously experiment. What works is repeated, and what doesn't is discarded. Coordinated experimentation enables ideas to be systematically tested and recorded, creating an institutional knowledge base on experimentation that can fuel what ideas are promoted to the entire faculty. Who have you turned to for learning about emerging practices? Consider formalizing a group of explorers who can test out ideas and feel part of a like-minded cohort.

For Technologists

The following are some considerations for those working in the technology industry related to AI for teaching assistance.

Design for customization. If a technology aims to enhance human relationships, it must be adaptable to people rather than requiring people to adapt to it. For the classroom, where every teacher and student is different, rather than designing a tool to be used in specific ways, thereby forcing the people to adapt to how the technology functions, design tools that teachers can use in many ways. For AI teaching assistants, scaffold interactivity with the teacher using your product so they can refine what the product does. Let them talk to it, focus its attention, and choose how and on what they would prefer to receive feedback.

Connect outputs from other AI assistants as inputs for the AI teaching assistant. Just as team teaching is best when teachers have the opportunity to plan together, so will a teaching assistant be best if it is connected to lesson objectives and a larger unit plan. When a class veers in an unanticipated direction—which happens for valid reasons related to learning—rather than seeking to correct the direction of the class, AI could help the teacher integrate the new information into the homework plan or propose to the teacher a revised end of class. Similarly, if an AI teaching assistant is drawing on students' strengths and weaknesses from its evening sessions with students, it can best support teachers' actions live in class, offering examples during a class digression that speak to anticipated student needs. Wiring in information from the planning assistant, learning assistant, and doing assistant will provide the most useful and purposeful support for teachers.

Design outputs from the AI teaching assistant to be inputs for other AI assistants. Information from class must be connected back to the planning assistant and to the student-facing learning and doing assistants. For teachers, necessary adjustments may be recommended and made to a lesson plan based on what transpired in class. For students, materials covered in class, including specific guidelines recommended by teachers, can support individual student mastery and productivity when they work on their own outside of class. Wiring information out from the teaching assistant to the other assistants ensures continuity of experience, both inside and outside of the classroom.

Epilogue

To Reach a Positive Alternative Future, We Must Imagine It First

We have Paleolithic emotions,
medieval institutions and godlike technology.

—EDWARD O. WILSON

We hope that this book has contributed to imagining a positive alternative future for AI and schools. The future we explored in this book is one that, for some colleagues, may be too radical, and for others too modest. For us, it is a future in which our increasingly extraordinary technological creations serve and support our unchanging human nature. It has always been so. From writing and the printing press to the dictionary, the calculator, the word processor, and the internet, new media have repeatedly transformed elements of the learning experience, but they have done so without rewriting core elements of education. In Socrates's time, students of the School of Athens met together with their teacher, just as teachers and students do today and will continue to in the future.

AI is the latest step in our human technological tradition. We hope that the imagined future we present is one that propels discourse at your school past fears and toward topics that are essential to today's youth, like reclaiming agency; fostering taste, trust, and teamwork; and more. While you engage in this discourse with colleagues and students, we hope that some of the frameworks we have examined in this book fuel both your questions and your confidence.

The More Things Change . . .

At a doctor's appointment, Peter's physician asked if she could use AI to listen to and summarize the appointment. At the beginning of the check-up, she opened an app on her phone, and the two of them talked through various routine questions about symptom A, habit B, and so on. They were able to speak uninterrupted. She was able to focus her full attention on the conversation and never broke concentration to take notes. Near the end of the appointment, they together reviewed AI's notes on a desktop computer, which had automatically synchronized information from the phone app. The notes accurately represented the medical components of the conversation, omitted tangents and irrelevant details, and identified follow-up items. The record required no alteration for accuracy. Peter and his doctor were able to communicate, uninhibited by the need to take notes. This meant they could sit face to face, the doctor could more readily follow up on key pieces of information, and Peter could more easily feel seen and heard. The AI assistant in the medical appointment was a small, prosocial improvement that freed the doctor's appointment to be more human centric.

. . . The More They Stay the Same

What didn't change in this arrangement was a doctor meeting with a patient to talk about the patient's experiences since the last check-up. What didn't change was the doctor weighing which next steps for her patient's optimal health should be based on medical interventions or on recommendations for personal habits or lifestyle changes. These are moments of human inquiry and judgment that are based on—and contribute to—the human relationship in the room. While AI may better read a radiology scan than a human and better filter through thousands of academic studies for a specific medical approach, humans—for now—are still necessary to read a room or to build a connection that might more readily change a patient's mind. This is evolving, as we've explored here, and AI will begin picking up some of this social and emotional work—but with life's infinite variability and with the impossibility of digitizing all that affects our human experience, humans will still be needed to ground the meaning and significance of what happens in our lives.

Medicine and education are on similar trajectories. In both, growing portions of our work can be supported by or delegated to machines. Also in both, some elements of our human relationships and agency remain central to what we do. Some parts of this work, in both contexts, won't change.

In education, what won't change is the need for teachers and students to come together, for students to learn and practice on their own outside of class, and for teachers to reflect each day on what might be best for those under their care. But what *will* change is that AI will increasingly play a role in supporting all three of these experiences—and students and teachers will increasingly need to engage these moments with agency. AI's reach will grow. It will help students develop content mastery, offer feedback on work, suggest college pathways, and maybe even provide companionship.

And it will keep evolving. Just as teachers today have countless diverse approaches to teaching grammar, algebra, and world history, so, too, will countless AI apps, tools, and new approaches emerge to further support mastery of specific disciplines and skills. These new apps, tools, and approaches will inevitably be more interactive, multimedia rich, conversant, and full of feedback opportunities. And however powerful they become, they will eventually be surpassed again by future technologies, beyond what we imagine today.

Yet, as we wake each morning and come together in school, the *whys* of what we do will not change. We will work to deepen our relationships in our communities, whether that's by serving academic, humanist, social, or pragmatic teaching philosophies. And in doing this, we will need to critically assess the degree to which the technology that has entered further and further into our lives serves or does a disservice to our inherently social educational imperative.

Our shared meaning making will continue to take place at the messy speed of humans. Our students—and our teachers—will struggle then as they do now with forming the right questions, evaluating what they learn, and applying their values to their lives. We may practice these skills and habits with our AI tools, but we will test them, grow them, and use them with each other.

AI assistants are changing how we research, plan, learn, do, and give and receive feedback, but automation's extension of our human capabilities means that human decision making and human relationships are more important, not less. If we remember this, then regardless of what technology next comes our way, we can know that our agentic and prosocial purposes are—and must continue to be—at the core of what we do in schools for as long as humans live together in a free and civil society.

References and Resources

Abry, T., Hulleman, C. S., & Rimm-Kaufman, S. E. (2015). Using indices of fidelity to intervention core components to identify program active ingredients. *American Journal of Evaluation, 36*(3), 320–338.

Ackermann, H., Henke, A., Chevalère, J., Yun, H. S., Hafner, V. V., Pinkwart, N., et al. (2025). Physical embodiment and anthropomorphism of AI tutors and their role in student enjoyment and performance. *NPJ Science of Learning, 10*(1), Article 1.

The Age. (2007, September 13). *The best is yet to come.* Accessed at www.theage.com.au/business/small-business/the-best-is-yet-to-come-20090619-co4x.html on April 11, 2025.

Alkhatlan, A., & Kalita, J. (2019). Intelligent tutoring systems: A comprehensive historical survey with recent developments. *International Journal of Computer Applications, 181*(43).

Almegren, A., Mahdi, H. S., Hazaea, A. N., Ali, J. K., & Almegren, R. M. (2024). Evaluating the quality of AI feedback: A comparative study of AI and human essay grading. *Innovations in Education and Teaching International.* Advance online publication.

Alpert, A., Buizer, J. L., Cook, J., Dave, A., Furlow, J., Preston, K., et al. (2018). Climate effects on U.S. international interests. In D. R. Reidmiller, C. W. Avery, D. R. Easterling, K. E. Kunkel, K. L. M. Lewis, T. K. Maycock, et al. (Eds.), *Impacts, risks, and adaptation in the United States: Fourth National Climate Assessment, volume II* (pp. 595–628). U.S. Government Publishing Office.

American National Standards Institute. (1984). *Specifications for making buildings and facilities accessible to, and usable by, the physically handicapped.* Author. Accessed at https://education.ohio.gov/getattachment/Topics/Career-Tech/Civil-Rights-Methods-of-Administration-Program/ANSI.pdf.aspx on April 10, 2025.

An, Y., Yu, J. H., & James, S. (2025). Investigating the higher education institutions' guidelines and policies regarding the use of generative AI in teaching, learning, research, and administration. *International Journal of Educational Technology in Higher Education, 22*, Article 10.

Anderson, J. R., Corbett, A. T., Koedinger, K. R., & Pelletier, R. (1995). Cognitive tutors: Lessons learned. The journal of the learning sciences, 4(2), 167–207.

Anderson, K. R. (2008, March 18). *Co-teaching: A literature review.* Author. Accessed at https://education.wm.edu/centers/ttac/documents/articles/coteachingliteraturereview.pdf on April 10, 2025.

Artman-Meeker, K., Rosenberg, N., Badgett, N., Yang, X., & Penney, A. (2017). The effects of bug-in-ear coaching on pre-service behavior analysts' use of functional communication training. *Behavior Analysis in Practice, 10*(3), 228–241. https://doi.org/10.1007/s40617-016-0166-4

Bacon, D. R. (2005). The effect of group projects on content-related learning. *Journal of Management Education, 29*(2), 248–267.

Baddeley, A. (2003). Working memory: Looking back and looking forward. *Nature Reviews Neuroscience, 4*(10), 829–839.

Baker, H. (2021, July 10). *How many atoms are in the observable universe?* Live Science. Accessed at www.livescience.com/how-many-atoms-in-universe.html on April 10, 2025.

Bansal, G., Nushi, B., Kamar, E., Horvitz, E., & Weld, D. S. (2021). Is the most accurate AI the best teammate? Optimizing AI for teamwork. *Proceedings of the AAAI Conference on Artificial Intelligence, 35*(13), 11405–11414.

Barnett, C. (1960). *The desert generals.* Viking Press.

Barr, A., Beard, M., & Atkinson, R. C. (1976). The computer as a tutorial laboratory: the Stanford BIP project. *International Journal of Man-Machine Studies, 8*(5), 567–596.

Bastani, H., Bastani, O., Sungu, A., Ge, H., Kabakcı, Ö., & Mariman, R. (2025). Generative AI without guardrails can harm learning: Evidence from high school mathematics. *PNAS, 122*(26), Article e2422633122.

Beer, S. (1985). *Diagnosing the system for organizations.* Wiley.

Belpaeme, T., Kennedy, J., Ramachandran, A., Scassellati, B., & Tanaka, F. (2018). Social robots for education: A review. *Science Robotics, 3*(21).

Bjork, R. A. (1994). Memory and metamemory considerations in the training of human beings. In J. Metcalfe & A. P. Shimamura (Eds.), *Metacognition: Knowing about knowing* (pp. 185–205). The MIT Press.

Blatchford, P., Bassett, P., Brown, P., Martin, C., Russell, A., & Webster, R. (2009, August). *Deployment and impact of support staff in schools: Characteristics, working conditions and job satisfaction of support staff in schools (strand 1, waves 1–3 in 2004, 2006 and 2008)* (Research Report No. DCSF-RR154). Department for Children, Schools and Families. Accessed at https://discovery.ucl.ac.uk/id/eprint/10096852/1/Blatchford_DISS_project_S1W123_final_report.pdf on April 10, 2025.

Blodgett, S. L., Barocas, S., Daumé, H., III, & Wallach, H. (2020). Language (technology) is power: A critical survey of "bias" in NLP. In D. Jurafsky, J. Chai, N. Schluter, & J. Tetreault (Eds.), *Proceedings of the 58th Annual Meeting of the Association for Computational Linguistics* (pp. 5454–5476). Association for Computational Linguistics.

Bloom, B. S., Engelhart, M. D., Furst, E. J., Hill, W. H., & Krathwohl, D. R. (1956). *Taxonomy of educational objectives: The classification of educational goals—Handbook 1. Cognitive domain.* Longmans.

Bolukbasi, T., Chang, K.-W., Zou, J., Saligrama, V., & Kalai, A. (2016). Man is to computer programmer as woman is to homemaker? Debiasing word embeddings. In D. D. Lee & U. von Luxburg (Eds.), *Proceedings of the 30th International Conference on Neural Information Processing Systems* (pp. 4356–4364). Curran Associates.

Bozkurt, A., Xiao, J., Farrow, R., Bai, J. Y. H., Nerantzi, C., Moore, S., et al. (2024). The manifesto for teaching and learning in a time of generative AI: A critical collective stance to better navigate the future. *Open Praxis, 16*(4), 487–513. https://doi.org/10.55982/openpraxis.16.4.777

Bransford, J. D., Brown, A. L., & Cocking, R. R. (Eds.). (2000). *How people learn: Brain, mind, experience, and school* (Expanded ed.). National Academies Press.

Bregman, R. (2025). *Moral ambition: Stop wasting your talent and start making a difference* (E. Moore, Trans.). Little, Brown.

Brown, T. (2009). *Change by design: How design thinking transforms organizations and inspires innovation.* Harper Business.

Bryant, J., Heitz, C., Sanghvi, S., & Wagle, D. (2020, January 14). *How artificial intelligence will impact K–12 teachers.* Accessed at www.mckinsey.com/industries/education/our-insights/how-artificial-intelligence-will-impact-k-12-teachers on April 10, 2025.

Buolamwini, J., & Gebru, T. (2018). Gender shades: Intersectional accuracy disparities in commercial gender classification. In S. A. Friedler & C. Wilson (Eds.), *Proceedings of Machine Learning Research: Volume 81. Conference on Fairness, Accountability and Transparency* (pp. 77–91).

Cabi, E. (2018). The impact of the flipped classroom model on students' academic achievement. *International Review of Research in Open and Distributed Learning, 19*(3), 202–221.

Calarco, J. M. (2018). *Negotiating opportunities: How the middle class secures advantages in school.* Oxford University Press.

Cao, S., Jiang, W., Wang, J., & Yang, B. (2024). From man vs. machine to man + machine: The art and AI of stock analyses. *Journal of Financial Economics, 160,* Article 103910.

Carbonell, J. R. (1970). AI in CAI: An artificial-intelligence approach to computer-assisted instruction. *IEEE Transactions on Man-Machine Systems, 11*(4).

Cardona, M. A., Rodríguez, R. J., & Ishmael, K. (2023, May). *Artificial intelligence and the future of teaching and learning: Insights and recommendations.* Office of Educational Technology. Accessed at www.ed.gov/sites/ed/files/documents/ai-report/ai-report.pdf on April 10, 2025.

Carnegie Learning. (2025). *The state of AI in education 2025: Key findings from a national survey.* Author. Accessed at https://discover.carnegielearning.com/hubfs/PDFs/Whitepaper%20and%20Guide%20PDFs/2025-AI-in-Ed-Report.pdf?hsLang=en on September 25, 2025.

Casey, B. J., Getz, S., & Galvan, A. (2008). The adolescent brain. *Developmental Review, 28*(1), 62–77.

CAST. (2008). *Universal Design for Learning Guidelines version 1.0.* Author. Accessed at https://udlguidelines.cast.org/static/udlg_graphicorganizer_v1-0.pdf on April 10, 2025.

CAST. (2024). *About the Guidelines 3.0 update.* Accessed at https://udlguidelines.cast.org/more/about-guidelines-3-0 on April 11, 2025.

Center on the Developing Child. (n.d.). *Executive function: Skills for life and learning.* Author. Accessed at https://developingchild.harvard.edu/wp-content/uploads/2024/10/InBrief-Executive-Function-Skills-for-Life-and-Learning-2.pdf on September 16, 2025.

Chai, F., Ma, J., Wang, Y., Zhu, J., & Han, T. (2024). Grading by AI makes me feel fairer? How different evaluators affect college students' perception of fairness. *Frontiers in Psychology, 15,* Article 1221177.

Chang, K. (2011, January 17). Bending and stretching classroom lessons to make math inspire. *The New York Times.* Accessed at www.nytimes.com/2011/01/18/science/18prof.html on April 10, 2025.

Chein, J., Albert, D., O'Brien, L., Uckert, K., & Steinberg, L. (2011). Peers increase adolescent risk taking by enhancing activity in the brain's reward circuitry. *Developmental Science, 14*(2), F1–F10.

Cheung, D. (2000). Measuring teachers' meta-orientations to curriculum: Application of hierarchical confirmatory factor analysis. *The Journal of Experimental Education, 68*(2), 149–165. https://doi.org/10.1080/00220970009598500

Childs, T. M., & Wooten, N. R. (2023). Teacher bias matters: An integrative review of correlates, mechanisms, and consequences. *Race Ethnicity and Education, 26*(3), 368–397.

Chitiyo, J. (2017). Challenges to the use of co-teaching by teachers. *International Journal of Whole Schooling, 13*(3), 55–66.

Cicero. (1942). *On the orator: Books 1–2* (E. W. Sutton & H. Rackham, Trans.). Harvard University Press. (Original work 55 BCE)

Clark, A., & Chalmers, D. (1998). The extended mind. *Analysis, 58*(1), 7–19.

Cook, L., & Friend, M. (1995). Co-teaching: Guidelines for creating effective practices. *Focus on Exceptional Children, 28*(3), 1–16.

Cookson, P. W., Jr., & Persell, C. H. (1985). *Preparing for power: America's elite boarding schools.* Basic Books.

Costello, T. H., Pennycook, G., & Rand, D. G. (2024). Durably reducing conspiracy beliefs through dialogues with AI. *Science, 385*(6714).

Cotton, D. R. E., Cotton, P. A., & Shipway, J. R. (2024). Chatting and cheating: Ensuring academic integrity in the era of ChatGPT. *Innovations in Education and Teaching International, 61*(2), 228–239.

Craft, A. (2005). *Creativity in schools: Tensions and dilemmas.* Routledge.

Criss, C. J., Konrad, M., Alber-Morgan, S. R., Brock, M. E., & Harris, A. B. (2024). Using performance feedback with and without goal setting on teachers' classroom management skills. *Behavioral Disorders, 49*(2), 91–105.

Csikszentmihalyi, M. (2000). Happiness, flow, and economic equality. *American Psychologist, 55*(10), 1163–1164.

Cuban, L. (2012, January 14). *The multi-layered curriculum: Why change is often confused with reform* [Blog post]. Accessed at https://nepc.colorado.edu/blog/multi-layered-curriculum on April 10, 2025.

Dai, W., Lin, J., Jin, H., Li, T., Tsai, Y.-S., Gašević, D., et al. (2023). Can large language models provide feedback to students? A case study on ChatGPT. In M. Chang, N.-S. Chen, R. Kuo, G. Rudolph, D. G. Sampson, & A. Tlili (Eds.), *2023 IEEE International Conference on Advanced Learning Technologies* (pp. 323–325).

Dalalah, D., & Dalalah, O. M. A. (2023). The false positives and false negatives of generative AI detection tools in education and academic research: The case of ChatGPT. *The International Journal of Management Education, 21*(2), Article 100822.

Darling-Hammond, L., Wei, R. C., Andree, A., Richardson, N., & Orphanos, S. (2009, February). *Professional learning in the learning profession: A status report on teacher development in the United States and abroad.* National Staff Development Council. Accessed at https://edpolicy.stanford.edu/sites/default/files/publications/professional-learning-learning-profession-status-report-teacher-development-us-and-abroad_0.pdf on April 10, 2025.

Davis, R. O., Park, T., & Vincent, J. (2021). A systematic narrative review of agent persona on learning outcomes and design variables to enhance personification. *Journal of Research on Technology in Education, 53*(1), 89–106.

De Backer, L., Schelfhout, W., Simons, M., Vandervieren, E., & Rivera Espejo, J. (2023). Impact of team teaching on student teachers' professional identity: A Bayesian approach. *Education Sciences, 13*(11), Article 1087.

Deci, E. L., & Ryan, R. M. (1985). The general causality orientations scale: Self-determination in personality. *Journal of Research in Personality, 19*(2), 109–134.

Dedehayir, O., & Steinert, M. (2016). The hype cycle model: A review and future directions. *Technological Forecasting and Social Change, 108*, 28–41.

Dee, T., & Gershenson, S. (2017). *Unconscious bias in the classroom: Evidence and opportunities.* Google. Accessed at https://services.google.com/fh/files/misc/unconscious-bias-in-the-classroom-report.pdf on April 10, 2025.

Demetriou, A., Spanoudis, G., & Mouyi, A. (2011). Educating the developing mind: Towards an overarching paradigm. *Educational Psychology Review, 23*(4), 601–663.

Deresiewicz, W. (2014). *Excellent sheep: The miseducation of the American elite and the way to a meaningful life.* Free Press.

Desimone, L. M. (2011). A primer on effective professional development. *Phi Delta Kappan, 92*(6), 68–71.

Dewey, J. (1938). *Experience and education.* Macmillan.

Di Canio, D. (2021). *The impact of robot gestures on student reasoning about geometrical conjectures* [Unpublished master's thesis]. University of Illinois at Chicago.

Diamond, A. (2013). Executive functions. *Annual Review of Psychology, 64*, 135–168.

Dickson, B., & Housiaux, A. (n.d.). *Feedback in practice: Research for teachers.* Tang Institute at Andover. Accessed at https://d3g5kqt8mzsgiz.cloudfront.net/images/File/Feedback-in-Practice.pdf on April 10, 2025.

Diliberti, M. K., Schwartz, H. L., Doan, S., Shapiro, A., Rainey, L. R., & Lake, R. J. (2024). *Using artificial intelligence tools in K–12 classrooms* (RR-A956-21). RAND. Accessed at www.rand.org/content/dam/rand/pubs/research_reports/RRA900/RRA956-21/RAND_RRA956-21.pdf on April 10, 2025.

D'Mello, S., & Graesser, A. (2012). Dynamics of affective states during complex learning. *Learning and Instruction*, *22*(2), 145–157.

D'Mello, S., & Graesser, A. (2013). AutoTutor and affective autotutor: Learning by talking with cognitively and emotionally intelligent computers that talk back. *ACM Transactions on Interactive Intelligent Systems*, *2*(4), Article 23.

Dong, A., Jong, M. S.-Y., & King, R. B. (2020). How does prior knowledge influence learning engagement? The mediating roles of cognitive load and help-seeking. *Frontiers in Psychology*, *11*, Article 591203.

Doyle, T. (2011). *Learner-centered teaching: Putting the research on learning into practice*. Stylus.

Drucker, P. F. (2008). *Management* (Revised ed.). Harper Business.

Dunlosky, J., Rawson, K. A., Marsh, E. J., Nathan, M. J., & Willingham, D. T. (2013). Improving students' learning with effective learning techniques: Promising directions from cognitive and educational psychology. *Psychological Science in the Public Interest*, *14*(1), 4–58.

Dzieza, J. (2024, December 3). *Friend or faux?* The Verge. Accessed at www.theverge.com/c/24300623/ai-companions-replika-openai-chatgpt-assistant-romance on April 10, 2025.

Eccles, J. S., Midgley, C., Wigfield, A., Buchanan, C. M., Reuman, D., Flanagan, C., et al. (1993). Development during adolescence: The impact of stage-environment fit on young adolescents' experiences in schools and in families. *American Psychologist*, *48*(2), 90–101.

EdWeek Research Center. (n.d.). *1st annual Merrimack College Teacher Survey: 2022 results*. Author. Accessed at https://guinote.wordpress.com/wp-content/uploads/2024/02/merrimack_teachers_are_deeply_disillusioned_survey_data_confirms.pdf on April 10, 2025.

Edyburn, D. (2003). Research and practice. *Journal of Special Education Technology*, *19*(1), 60–64. https://doi.org/10.1177/016264340401900106

Ehri, L. C., Nunes, S. R., Stahl, S. A., & Willows, D. M. (2001). Systematic phonics instruction helps students learn to read: Evidence from the National Reading Panel's meta-analysis. *Review of Educational Research*, *71*(3), 393–447.

Eikeland, I., & Ohna, S. E. (2022). Differentiation in education: A configurative review. *Nordic Journal of Studies in Educational Policy*, *8*(3), 157–170.

Eisenhower, D. (1984). *The papers of Dwight David Eisenhower* (Vol. 11; L. Galambos, Ed.). Johns Hopkins Press.

Eisner, E. W., & Vallance, E. (Eds.). (1974). *Conflicting conceptions of curriculum*. McCutchan.

Elkhatat, A. M., Elsaid, K., & Almeer, S. (2023). Evaluating the efficacy of AI content detection tools in differentiating between human and AI-generated text. *International Journal for Educational Integrity*, *19*, Article 17.

Ellington, A. J. (2003). A meta-analysis of the effects of calculators on students' achievement and attitude levels in precollege mathematics classes. *Journal for Research in Mathematics Education*, *34*(5), 433–463.

Emberton, M. (2021). Unconscious bias is a human condition. *The Permanente Journal*, *25*(2).

Escalante, J., Pack, A., & Barrett, A. (2023). AI-generated feedback on writing: Insights into efficacy and ENL student preference. *International Journal of Educational Technology in Higher Education*, *20*, Article 57.

Facer, K. (2011). *Learning futures: Education, technology and social change*. Routledge.

Fadel, C., Bialik, M., & Trilling, B. (2015). *Four-dimensional education: The competencies learners need to succeed.* Center for Curriculum Redesign.

Fadel, C., Black, A., Taylor, R., Slesinski, J., & Dunn, K. (2024). *Education for the age of AI*. Center for Curriculum Redesign.

Faggioli, G., Dietz, L., Clarke, C. L. A., Demartini, G., Hagen, M., Hauff, C., et al. (2024). Who determines what is relevant? Humans or AI? Why not both? *Communications of the ACM*, *67*(4), 31–34. https://doi.org/10.1145/3624730

Feldon, D. F. (2007). Cognitive load and classroom teaching: The double-edged sword of automaticity. *Educational Psychologist*, *42*(3), 123–137.

Firmender, J. M., Reis, S. M., & Sweeny, S. M. (2013). Reading comprehension and fluency levels ranges across diverse classrooms: The need for differentiated reading instruction and content. *Gifted Child Quarterly*, *57*(1), 3–14.

Foer, J. (2011). *Moonwalking with Einstein: The art and science of remembering everything*. Penguin Books.

Friend, M., & Cook, L. (1992). *Interactions: Collaboration skills for school professionals*. Longman.

Fuller, R. (2013, December 18). *Teaching isn't rocket science. It's harder*. Slate. Accessed at https://slate.com/human-interest/2013/12/teaching-in-americas-highest-need-communities-isnt-rocket-science-its-harder.html on April 11, 2025.

Gallup. (2014). *Great jobs great lives: The 2014 Gallup-Purdue index report*. Author. Accessed at www.luminafoundation.org/wp-content/uploads/2017/08/galluppurdueindex-report-2014.pdf on April 11, 2025.

Gallup & Walton Family Foundation. (2024). *Walton Family Foundation-Gallup Voices of Gen Z study: Year 2 annual survey report*. Authors. Accessed at https://nextgeninsights.waltonfamilyfoundation.org/wp-content/uploads/2024/08/Walton_Gallup_Voices-of-Gen-Z_Year-2-2024-Final-Report.pdf on April 11, 2025.

Geary, D. C. (2011). Cognitive predictors of achievement growth in mathematics: A 5-year longitudinal study. *Developmental Psychology*, *47*(6), 1539–1552.

Gehlert, T., Schulze, C., & Schlag, B. (2012). Evaluation of different types of dynamic speed display signs. *Transportation Research Part F: Traffic Psychology and Behaviour*, *15*(6), 667–675.

Gess-Newsome, J. (1999). Pedagogical content knowledge: An introduction and orientation. In J. Gess-Newsome & N. G. Lederman (Eds.), *Examining pedagogical content knowledge* (pp. 3–20). Kluwer Academic.

Gibran, K. (1923). *The prophet*. Knopf.

Gilal, F. G., Zhang, J., Gilal, N. G., & Gilal, R. G. (2018). Association between a parent's brand passion and a child's brand passion: A moderated moderated-mediation model. *Psychology Research and Behavior Management*, *11*, 91–102.

Gleick, J. (2011). *The information: A history, a theory, a flood.* Vintage Books.

Global Silicon Valley. (2024, April 30). *The difference between great AI and great teaching with Dan Meyer | ASU+GSV SHOW* [Video file]. Accessed at www.youtube.com/watch?v=iH4Pn4bpOfQ on April 11, 2025.

Goetz, T. (2011, June 19). Harnessing the power of feedback loops. *Wired.* Accessed at www.wired.com/2011/06/ff-feedbackloop on April 11, 2025.

Golann, J. W. (2015). The paradox of success at a no-excuses school. *Sociology of Education, 88*(2), 103–119.

Goodlad, J. I. (1979). *Curriculum inquiry: The study of curriculum practice.* McGraw-Hill.

Google Books Ngram Viewer. (n.d.). *Instructional design.* Accessed at https://books.google.com/ngrams/graph?content=instructional+design&year_start=1800&year_end=2022&corpus=en&smoothing=3 on April 11, 2025.

Google & Ipsos. (2025, January). *Our life with AI: From innovation to application.* Authors. Accessed at https://static.googleusercontent.com/media/publicpolicy.google/en//resources/ipsos_google_our-life-with-ai_2024_25.pdf on April 11, 2025.

Graesser, A. C., McNamara, D. S., Louwerse, M. M., & Cai, Z. (2004). Coh-metrix: Analysis of text on cohesion and language. *Behavior Research Methods, Instruments, and Computers, 36*(2), 193–202.

Greenwald, A. G., & Banaji, M. R. (1995). Implicit social cognition: Attitudes, self-esteem, and stereotypes. *Psychological Review, 102*(1), 4–27.

Greenwald, A. G., McGhee, D. E., & Schwartz, J. L. (1998). Measuring individual differences in implicit cognition: The implicit association test. *Journal of Personality and Social Psychology, 74*(6), 1464–1480.

Griffin, W. G., & Morrison, D. (2010). *The creative process illustrated: How advertising's big ideas are born.* HOW Books.

Guilford, J. P. (1956). The structure of intellect. *Psychological Bulletin, 53*(4), 267–293. https://doi.org/10.1037/h0040755

Guzik, E. E., Byrge, C., & Gilde, C. (2023). The originality of machines: AI takes the Torrance Test. *Journal of Creativity, 33*(3), Article 100065.

Hamblin, L., Barker, D., & Arghode, V. (2020). A phenomenological approach to explore faculty perceptions about invisible labor. *Community College Journal of Research and Practice, 44*(10–12), 804–818.

Hamilton, E. R., Rosenberg, J. M., & Akcaoglu, M. (2016). The substitution augmentation modification redefinition (SAMR) model: A critical review and suggestions for its use. *TechTrends, 60*(5), 433–441.

Hansard HC Deb. (28 October 1943). vol. 393, col. 403.

Happy, S. L., Dasgupta, A., Patnaik, P., & Routray, A. (2013). Automated alertness and emotion detection for empathic feedback during e-Learning. In Kinshuk & Sridhar Iyer (Eds.), *2013 IEEE Fifth International Conference on Technology for Education* (pp. 47-50). IEEE.

Hart, M. (1990). Critical theory and beyond: Further perspectives on emancipatory education. *Adult Education Quarterly, 40*(3), 125–138.

Hattie, J. (2012). *Visible learning for teachers: Maximizing impact on learning*. Routledge.

Hattie, J., & Timperley, H. (2007). The power of feedback. *Review of Educational Research, 77*(1), 81–112.

Hau, I. C. (2025). *Love to learn: The transformative power of care and connection in early education*. PublicAffairs.

Hembree, R., & Dessart, D. J. (1986). Effects of hand-held calculators in precollege mathematics education: A meta-analysis. *Journal for Research in Mathematics Education, 17*(2), 83–99.

Hill, K. (2025, January 15). She is in love with ChatGPT. *The New York Times*. Accessed at www.nytimes.com/2025/01/15/technology/ai-chatgpt-boyfriend-companion.html on April 11, 2025.

Hitchcock, C., Meyer, A., Rose, D., & Jackson, R. (2002). Providing new access to the general curriculum: Universal Design for Learning. *TEACHING Exceptional Children, 35*(2), 8–17.

Hlado, P., & Harvankova, K. (2024). Teachers' perceived work ability: A qualitative exploration using the Job Demands-Resources model. *Humanities and Social Sciences Communications, 11*, Article 304.

Hlado, P., Juhaňák, L., & Harvankova, K. (2025). Antecedents of perceived teacher work ability: A comprehensive model across work and non-work domains. *Frontiers in Psychology, 16*, Article 1557456.

Ho, V. T., Wong, S.-S., & Lee, C. H. (2011). A tale of passion: Linking job passion and cognitive engagement to employee work performance. *Journal of Management Studies, 48*(1), 26–47.

Hoffman, R., & Beato, G. (2025). *Superagency: What could possibly go right with our AI future*. Authors Equity.

Hohenstein, J., & Jung, M. (2020). AI as a moral crumple zone: The effects of AI-mediated communication on attribution and trust. *Computers in Human Behavior, 106*, Article 106190.

Hollett, N. L., Brock, S. J., & Hinton, V. (2017). Bug-in-ear technology to enhance preservice teacher training: Peer versus instructor feedback. *International Journal of Learning, Teaching and Educational Research, 16*(2), 1–10.

Holmes, W., Bialik, M., & Fadel, C. (2019). *Artificial intelligence in education: Promises and implications for teaching and learning*. Center for Curriculum Redesign.

Holmes, W., Porayska-Pomsta, K., Holstein, K., Sutherland, E., Baker, T., Shum, S. B., et al. (2022). Ethics of AI in education: Towards a community-wide framework. *International Journal of Artificial Intelligence in Education, 32*(3), 504–526.

Holmes, W., & Tuomi, I. (2022). State of the art and practice in AI in education. *European Journal of Education, 57*(4), 542–570.

Holt, L. (2024, April 11). The 5 percent problem. *Education Next*. Accessed at www.educationnext.org/5-percent-problem-online-mathematics-programs-may-benefit-most-kids-who-need-it-least on April 11, 2025.

Hopmann, S. (2007). Restrained teaching: The Common Core of Didaktik. *European Educational Research Journal, 6*(2), 109–124.

Horn, A. L., Rock, M. L., Douglas, K. H., Bean, K. M., Layden, S. J., & Roitsch, J. (2022). Effects of teacher-delivered eCoaching on paraeducators and students with autism spectrum disorder. *Education and Training in Autism and Developmental Disabilities, 57*(3), 287–302.

Howard, S. K., & Mozejko, A. (2015). Teachers: Technology, change and resistance. In M. Henderson & G. Romeo (Eds.), *Teaching and digital technologies: Big issues and critical questions* (pp. 307–316). Cambridge University Press.

Hsu, H. (2025, June 30). What happens after A.I. destroys college writing? *The New Yorker*. Accessed at www.newyorker.com/magazine/2025/07/07/the-end-of-the-english-paper on September 17, 2025.

Immordino-Yang, M. H., & Damasio, A. (2007). We feel, therefore we learn: The relevance of affective and social neuroscience to education. *Mind, Brain, and Education, 1*(1), 3–10.

Ingersoll, R. M. (2009). *Who controls teachers' work?: Power and accountability in America's schools*. Harvard University Press.

Ipsos. (2022, January). *Global opinions and expectations about artificial intelligence: A global advisor survey*. Author. Accessed at www.ipsos.com/sites/default/files/ct/news/documents/2022-01/Global-opinions-and-expectations-about-AI-2022.pdf on April 11, 2025.

Ipsos. (2023, July). *Global views on A.I. 2023: How people across the world feel about artificial intelligence and expect it will impact their life*. Author. Accessed at www.ipsos.com/sites/default/files/ct/news/documents/2023-07/Ipsos%20Global%20AI%202023%20Report-WEB.pdf on April 11, 2025.

Jasanoff, S., & Kim, S.-H. (Eds.). (2015). *Dreamscapes of modernity: Sociotechnical imaginaries and the fabrication of power*. The University of Chicago Press.

Jenkins, S. B. (2009). Measuring teacher beliefs about curriculum orientations using the modified-curriculum orientations inventory. *The Curriculum Journal, 20*(2), 103–120. https://doi.org/10.1080/09585170902948798

John, P. D. (2006). Lesson planning and the student teacher: Re-thinking the dominant model. *Journal of Curriculum Studies, 38*(4), 483–498.

Johnston, C. M. (2021). *Creative Systems Theory: A comprehensive theory of purpose, change, and interrelationship in human systems (with particular pertinence to understanding the times we live in and the tasks ahead for the species)*. The Institute for Creative Development Press.

Johnston, S., Breeze, S., Hillard, R., Lewin, C., Nuttall, K., & O'Mahony, J. (2024, May 13). Generative AI in Asia Pacific: Young employees lead as employers play catch-up. *Deloitte Insights*. Accessed at www2.deloitte.com/us/en/insights/topics/emerging-technologies/generative-ai-adoption-asia-pacific-region.html?utm_source=chatgpt.com on April 11, 2025.

Juwah, C., Macfarlane-Dick, D., Matthew, B., Nicol, D., Ross, D., & Smith, B. (2004, June). *Enhancing student learning through effective formative feedback*. The Higher Education Academy. Accessed at www.enhancingfeedback.ed.ac.uk/documents/id353_senlef_guide.pdf on April 11, 2025.

Kalyuga, S., Ayres, P., Chandler, P., & Sweller, J. (2003). The expertise reversal effect. *Educational Psychologist, 38*(1), 23–31.

Kanat-Maymon, Y., Benjamin, M., Stavsky, A., Shoshani, A., & Roth, G. (2015). The role of basic need fulfillment in academic dishonesty: A self-determination theory perspective. *Contemporary Educational Psychology, 43*, 1–9.

Kang, H. (2017). Preservice teachers' learning to plan intellectually challenging tasks. *Journal of Teacher Education, 68*(1), 55–68.

Kang, S. K., DeCelles, K. A., Tilcsik, A., & Jun, S. (2016). Whitened résumés: Race and self-presentation in the labor market. *Administrative Science Quarterly, 61*(3), 469–502. https://doi.org/10.1177/0001839216639577

Kaplan, C., Chan, R., Farbman, D. A., & Novoryta, A. (2015). *Time for teachers: Leveraging expanded time to strengthen instruction and empower teachers*. National Center on Time & Learning. Accessed at https://files.eric.ed.gov/fulltext/ED561995.pdf on September 29, 2025.

Kasparov, G. (2010, February 11). The chess master and the computer. *The New York Review of Books*. Accessed at www.nybooks.com/articles/2010/02/11/the-chess-master-and-the-computer on April 11, 2025.

Kaufman, J. H., Doan, S., Tuma, A. P., Woo, A., Henry, D., & Lawrence, R. A. (2020). *How instructional materials are used and supported in U.S. K–12 classrooms: Findings from the 2019 American Instructional Resources Survey* (RR-A134-1). RAND Corporation. Accessed at www.rand.org/content/dam/rand/pubs/research_reports/RRA100/RRA134-1/RAND_RRA134-1.pdf on April 11, 2025.

Keppler, S., Sinchaisri, W. P., & Snyder, C. (2024). *Backwards planning with generative AI: Case study evidence from US K12 teachers*. SSRN. Accessed at https://papers.ssrn.com/sol3/papers.cfm?abstract_id=4924786 on April 11, 2025.

Kerr, S. T. (1981). How teachers design their materials: Implications for instructional design. *Instructional Science, 10*(4), 363–378

Khan, S. (2024). *Brave new words: How AI will revolutionize education (and why that's a good thing)*. Viking.

Kim, K.-N. (2019). Teachers' administrative workload crowding out instructional activities. *Asia Pacific Journal of Education, 39*(1), 31–49.

Kini, T., & Podolsky, A. (2016, June). *Does teaching experience increase teacher effectiveness? A review of the research*. Learning Policy Institute. Accessed at https://learningpolicyinstitute.org/sites/default/files/product-files/Teaching_Experience_Report_June_2016.pdf on September 19, 2025.

Kitchen, E., King, S. H., Robison, D. F., Sudweeks, R. R., Bradshaw, W. S., & Bell, J. D. (2006). Rethinking exams and letter grades: How much can teachers delegate to students? *CBE—Life Sciences Education, 5*(3), 270–280.

Kluger, A. N., & DeNisi, A. (1996). The effects of feedback interventions on performance: A historical review, a meta-analysis, and a preliminary feedback intervention theory. *Psychological Bulletin, 119*(2), 254–284.

Koedinger, K. R., Anderson, J. R., Hadley, W. H., & Mark, M. A. (1997). Intelligent tutoring goes to school in the big city. *International Journal of Artificial Intelligence in Education, 8*, 30–43.

Koehler, M. J., & Mishra, P. (2009). What is technological pedagogical content knowledge? *Contemporary Issues in Technology and Teacher Education, 9*(1), 60–70.

Kohlberg, L. (1981). *The philosophy of moral development: Moral stages and the idea of justice.*

Kotter, J. P. (1996). *Leading change.* Harvard Business School Press.

Krathwohl, D. R. (2002). A revision of Bloom's taxonomy: An overview. *Theory Into Practice, 41*(4), 212–218.

Krathwohl, D. R., Bloom, B. S., & Masia, B. B. (1964). *Taxonomy of educational objectives: The classification of educational goals—Handbook 2. Affective domain.* McKay.

Kulik, J. A., & Fletcher, J. D. (2016). Effectiveness of intelligent tutoring systems: A meta-analytic review. *Review of Educational Research, 86*(1), 42–78.

Kunar, M. A., Carter, R., Cohen, M., & Horowitz, T. S. (2008). Telephone conversation impairs sustained visual attention via a central bottleneck. *Psychonomic Bulletin and Review, 15*(6), 1135–1140.

Kurt, G., & Kurt, Y. (2024). Enhancing L2 writing skills: ChatGPT as an automated feedback tool. *Journal of Information Technology Education: Research, 23,* Article 24.

Lakoff, G., & Johnson, M. (1980). *Metaphors we live by.* The University of Chicago Press.

Laurillard, D. (2012). *Teaching as a design science: Building pedagogical patterns for learning and technology.* Routledge.

LearnPlatform. (n.d.). *EdTech top 40: A look at K–12 EdTech engagement during the 2023–24 school year.* Accessed at www.instructure.com/resources/research-reports/edtech-top-40-look-k-12-edtech-engagement-during-2023-24-school-year on April 11, 2025.

Lee, C. I. S., Peeters, D., Auer, J., Giesbers, B., & Appels, M. (2024). Grading and simultaneously providing high-information feedback: The Harmonized Appraisal assessment. In S. Taneja (Ed.), *Academy of Management proceedings.* Academy of Management.

Lee, K.-W., Mills, K., Ruiz, P., Coenraad, M., Fusco, J., Roschelle, J., et al. (2024, June 18). *AI literacy: A framework to understand, evaluate, and use emerging technology* [Blog post]. Accessed at https://digitalpromise.org/2024/06/18/ai-literacy-a-framework-to-understand-evaluate-and-use-emerging-technology on September 30, 2025.

Lee, V. R., Pope, D., Miles, S., & Zárate, R. C. (2024). Cheating in the age of generative AI: A high school survey study of cheating behaviors before and after the release of ChatGPT. *Computers and Education: Artificial Intelligence, 7,* Article 100253.

Lembke, A. (2021). *Dopamine nation: Finding balance in the age of indulgence.* Dutton.

Lemov, D. (2010). *Teach like a champion: 49 techniques that put students on the path to college, grades K–12.* Jossey-Bass.

Li, J., Jangamreddy, N. K., Hisamoto, R., Bhansali, R., Dyda, A., Zaphir, L., et al. (2024). AI-assisted marking: Functionality and limitations of ChatGPT in written assessment evaluation. *Australasian Journal of Educational Technology, 40*(4), 56–72.

Liang, W., Yuksekgonul, M., Mao, Y., Wu, E., & Zou, J. (2023). GPT detectors are biased against non-native English writers. *Patterns, 4*(7), Article 100779.

Lipnevich, A. A., & Smith, J. K. (2008). Response to assessment feedback: The effects of grades, praise, and source of information. *ETS Research Report Series, 2008*(1), i–57.

Locke, E. A., & Latham, G. P. (1990). Work motivation and satisfaction: Light at the end of the tunnel. *Psychological Science, 1*(4), 240–246.

Lovell, O. (2020). *Sweller's Cognitive Load Theory in action.* Catt Educational.

Lynch, S. (2022, October 17). *AI in the loop: Humans must remain in charge.* Stanford Institute for Human-Centered AI. Accessed at https://hai.stanford.edu/news/ai-loop-humans-must-remain-charge on September 23, 2025.

Ma, W., Adesope, O. O., Nesbit, J. C., & Liu, Q. (2014, June 23). Intelligent tutoring systems and learning outcomes: A meta-analysis. *Journal of Educational Psychology.* Advance online publication. http://dx.doi.org/10.1037/a0037123

Mace, R. L. (1985). Universal design: Barrier free environments for everyone. *Designers West, 33*(1), 147–152.

Maier, S. F., & Seligman, M. E. (1976). Learned helplessness: Theory and evidence. *Journal of Experimental Psychology: General, 105*(1), 3–46.

Maples, B., Cerit, M., Vishwanath, A., & Pea, R. (2024). Loneliness and suicide mitigation for students using GPT3-enabled chatbots. *npj Mental Health Research, 3,* Article 4. https://doi.org/10.1038/s44184-023-00047-6

Marche, S. (2022, December 6). The college essay is dead. *The Atlantic.* Accessed at www.theatlantic.com/technology/archive/2022/12/chatgpt-ai-writing-college-student-essays/672371 on September 30, 2025.

Martella, A. M., Schneider, D. W., O'Day, G. M., & Karpicke, J. D. (2024). Investigating the intensity and integration of active learning and lecture. *Journal of Applied Research in Memory and Cognition, 13*(3), 354–369. https://doi.org/10.1037/mac0000160

Maslej, N., Fattorini, L., Perrault, R., Parli, V., Reuel, A., Brynjolfsson, E., et al. (2024, April). *Artificial intelligence index report 2024.* Stanford Institute for Human-Centered AI. Accessed at https://hai-production.s3.amazonaws.com/files/hai_ai-index-report-2024-smaller2.pdf on April 11, 2025.

Mathews, H., & Cohen, J. (2022). *Explicit instruction: A brief review of what we know and next directions for research.* Routledge.

Mayer, M. (2006, May 17). *Creativity loves constraint* [Video file]. Accessed at https://stvp.stanford.edu/videos/creativity-loves-constraint on September 19, 2025.

McCutcheon, G. (1980). How do elementary school teachers plan? The nature of planning and influences on it. *The Elementary School Journal, 81*(1), 4–23.

McDonald, N., Johri, A., Ali, A., & Collier, A. H. (2025). Generative artificial intelligence in higher education: Evidence from an analysis of institutional policies and guidelines. *Computers in Human Behavior: Artificial Humans, 3,* Article 100121.

McKenney, S., Kali, Y., Markauskaite, L., & Voogt, J. (2015). Teacher design knowledge for technology enhanced learning: An ecological framework for investigating assets and needs. *Instructional Science, 43*(2), 181–202.

McLuhan, M. (1964). *Understanding media: The extensions of man.* McGraw-Hill.

McNeil, J. D. (1996). *Curriculum: A comprehensive introduction* (5th ed.). Wiley.

Mehrabian, A., & Ferris, S. R. (1967). Inference of attitudes from nonverbal communication in two channels. *Journal of Consulting Psychology, 31*(3), 248–252.

Mehrabian, A., & Wiener, M. (1967). Decoding of inconsistent communications. *Journal of Personality and Social Psychology, 6*(1), 109–114.

Merrimack College. (2022). *1st annual Merrimack College teacher survey: 2022 results.* EdWeek Research Center.

Meyer, D. (2010, March). *Math class needs a makeover* [Video file]. TED Conferences. Accessed at www.ted.com/talks/dan_meyer_math_class_needs_a_makeover on April 11, 2025.

Meyer, D. (2011). *Shower v bath* [Skit]. Accessed at https://mrmeyer.com/threeacts/showervbath on April 11, 2025.

Meyer, D. (2024, October 16). *AI's delivery problem & how tutor copilot solves it.* Accessed at https://danmeyer.substack.com/p/ais-delivery-problem-and-how-tutor on April 11, 2025.

Meyer, D. (2025, February 26). *How to waste a teacher.* Accessed at https://danmeyer.substack.com/p/how-to-waste-a-teacher on April 11, 2025.

Miao, F., & Cukurova, M. (2024). *AI competency framework for teachers.* UNESCO. https://doi.org/10.54675/ZJTE2084

Michigan Virtual. (n.d.). *Current-day examples of student AI use.* Accessed at https://michiganvirtual.org/resources/ai/student-use on April 11, 2025.

Microsoft. (2024, May 8). *AI at work is here. Now comes the hard part.* Authors. Accessed at https://assets-c4akfrf5b4d3f4b7.z01.azurefd.net/assets/2024/05/2024_Work_Trend_Index_Annual_Report_6_7_24_666b2e2fafceb.pdf on April 11, 2025.

Mintzberg, H. (1973). *The nature of managerial work.* Harper & Row.

Mishra, P., & Koehler, M. J. (2006). Technological pedagogical content knowledge: A framework for teacher knowledge. *Teachers College Record, 108*(6), 1017–1054.

Mollick, E. (2024). *Co-intelligence: Living and working with AI.* Portfolio/Penguin.

Monsell, S. (2003). Task switching. *Trends in Cognitive Sciences, 7*(3), 134–140.

Montessori, M. (1967). *The absorbent mind* (C. A. Claremont, Trans.). Holt, Rinehart and Winston. (Original work published 1949)

Moore, O. (2024, August 21). *The top 100 gen AI consumer apps: 3rd edition.* Accessed at https://a16z.com/100-gen-ai-apps-3 on April 10, 2025.

Morrone, M. (2024, July 24). *The AI boyfriend business is booming.* Axios. Accessed at www.axios.com/2024/07/24/ai-boyfriend-replika-nomi-chatbot on April 11, 2025.

Murawski, W. W., & Swanson, H. L. (2001). A meta-analysis of co-teaching research: Where are the data? *Remedial and Special Education, 22*(5), 258–267.

Murdoch, D., English, A. R., Hintz, A., & Tyson, K. (2020). Feeling heard: Inclusive education, transformative learning, and productive struggle. *Educational Theory, 70*(5), 653–679.

Murray, D. M. (2003). Teach writing as a process not product. In V. Villanueva (Ed.), *Cross-talk in comp theory: A reader* (2nd ed., pp. 3–6). National Council of Teachers of English.

Nagel, T. (1986). *The view from nowhere.* Oxford University Press.

National Center for Education Statistics. (2022, July 6). *More than 80 percent of U.S. public schools report pandemic has negatively impacted student behavior and socio-emotional development.* Accessed at https://nces.ed.gov/whatsnew/press_releases/07_06_2022.asp on April 11, 2025.

National Governors Association Center for Best Practices & Council of Chief State School Officers. (2010). *Common Core State Standards for English language arts & literacy in history/social studies, science, and technical subjects.* Authors. Accessed at https://learning.ccsso.org/wp-content/uploads/2022/11/ELA_Standards1.pdf on September 19, 2025.

Newton, D. P., & Newton, L. D. (2019). Humanoid robots as teachers and a proposed code of practice. *Frontiers in Education, 4*, Article 125.

NGSS Lead States. (2013). *Next Generation Science Standards: For states, by states.* National Academies Press.

Norman, D. A. (1999, May–June). Affordance, conventions, and design. *Interactions, 6*(3), 38–42.

OECD. (2023). *OECD employment outlook 2023.* Accessed at www.oecd.org/en/publications/oecd-employment-outlook-2023_08785bba-en.html on April 11, 2025.

Ong, W. J. (1982). *Orality and literacy: The technologizing of the word.* Methuen.

Ophir, E., Nass, C., & Wagner, A. D. (2009). Cognitive control in media multitaskers. *PNAS, 106*(37), 15583–15587.

Oregon State University Ecampus. (2024). *Bloom's taxonomy revisited.* Author. Accessed at https://ecampus.oregonstate.edu/faculty/artificial-intelligence-tools/blooms-taxonomy-revisited-v2-2024.pdf on September 30, 2025.

Overmier, J. B. (2002). On learned helplessness. *Integrative Physiological and Behavioral Science, 37*(1), 4–8.

Park, S., Takahashi, S., & White, T. (2014). *Developing an effective teacher feedback system (90-Day Cycle Report).* Carnegie Foundation for the Advancement of Teaching.

Parker, G. (2015). Teachers' autonomy. *Research in Education, 93*(1), 19–33.

Parnes, S. J. (1966). *Manual for institutes and programs.* Creative Education Foundation.

Patrick, S. K. (2022). Organizing schools for collaborative learning: School leadership and teachers' engagement in collaboration. *Educational Administration Quarterly, 58*(4), 638–673.

Paul, A. M. (2021). *The extended mind: The power of thinking outside the brain.* Houghton Mifflin Harcourt.

Pea, R. D. (1985). Beyond amplification: Using the computer to reorganize mental functioning. *Educational Psychologist, 20*(4), 167–182.

Perkins, M., Furze, L., Roe, J., & MacVaugh, J. (2024). The Artificial Intelligence Assessment Scale (AIAS) revisited: A framework for ethical integration of generative AI in educational assessment. *Journal of University Teaching and Learning Practice, 21*(6).

Perry, W. G., Jr. (1999). *Forms of intellectual and ethical development in the college years: A scheme.* Jossey-Bass.

Piaget, J. (1951). *Play, dreams and imitation in childhood* (C. Gattegno & F. M. Hodgson, Trans.). Routledge. (Original work published 1945)

Piaget, J. (1972). Intellectual evolution from adolescence to adulthood. *Human Development, 15*(1), 1–12.

Piaget, J., & Inhelder, B. (1969). *The psychology of the child* (H. Weaver, Trans.). Basic Books.

Pink, D. H. (2009). *Drive: The surprising truth about what motivates us.* Riverhead Books.

Plato. (2002). *Phaedrus* (R. Waterfield, Trans.). Oxford University Press. (Original work ca. 370 BCE)

Ploessl, D. M., & Rock, M. L. (2014). eCoaching: The effects on co-teachers' planning and instruction. *Teacher Education and Special Education, 37*(3), 191–215. https://doi.org/10.1177/0888406414525049

Polakova, P., & Ivenz, P. (2024). The impact of ChatGPT feedback on the development of EFL students' writing skills. *Cogent Education, 11*(1), Article 2410101.

Pratiwi, D. (2024). Pedagogical decisions in lesson planning and making use of the planning: Perspective of both experienced and novice teachers. *Journal of Educational Management and Instruction* (JEMIN), *4*, 213–224. https://doi.org/10.22515/jemin.v4i2.9497

Prosthesis. (n.d.). In *Merriam-Webster's online dictionary*. Accessed at www.merriam-webster.com/dictionary/prosthesis on April 11, 2025.

Prothero, A. (2024a, February 19). What is age-appropriate use of AI? 4 developmental stages to know about. *Education Week.* Accessed at www.edweek.org/technology/what-is-age-appropriate-use-of-ai-4-developmental-stages-to-know-about/2024/02 on September 30, 2025.

Prothero, A. (2024b, April 5). More teachers are using AI-detection tools. Here's why that might be a problem. *Education Week.* Accessed at www.edweek.org/technology/more-teachers-are-using-ai-detection-tools-heres-why-that-might-be-a-problem/2024/04 on April 11, 2025.

Quealy, K. (2017, November 14). The more education Republicans have, the less they tend to believe in climate change. *The New York Times.* Accessed at www.nytimes.com/interactive/2017/11/14/upshot/climate-change-by-education.html on April 11, 2025.

Ramírez Castañeda, V. (2025, January 21). *The creation of bad students: AI detection for non-native English speakers.* Accessed at https://dlab.berkeley.edu/news/creation-bad-students-ai-detection-non-native-english-speakers on April 11, 2025.

Rascoff, M. (2025, January 24). [Keynote address]. UNESCO International Day of Education 2025, "Artificial Intelligence and Education: Challenges and Opportunities," New York, NY, United States. Accessed at https://digitaleducation.stanford.edu/news/educational-ai-humanity-loop on September 30, 2025.

Recht, D. R., & Leslie, L. (1988). Effect of prior knowledge on good and poor readers' memory of text. *Journal of Educational Psychology, 80*(1), 16–20.

Reese, H. W. (2011). The learning-by-doing principle. *Behavioral Development Bulletin, 17*(1), 1–19.

Reich, J., & Dukes, J. (2024). *Toward a new theory of arrival technologies: The case of ChatGPT and the future of education technology after adoption.* EdArXiv. Accessed at https://osf.io/preprints/edarxiv/x6vn7_v1 on April 11, 2025.

Reich, J., & Ruipérez-Valiente, J. A. (2019). The MOOC pivot. *Science, 363*(6423), 130–131.

Resnick, M., & Silverman, B. (2005). Some reflections on designing construction kits for kids. In *Proceedings of the 2005 conference on Interaction Design and Children* (pp. 117–122). Association for Computing Machinery Press.

The Rithm Project. (2024, June 28). *Will AI strengthen or erode human-to-human relationships?* Accessed at www.medium.com/@info_42369/will-ai-strengthen-or-erode-human-to-human-relationships-4526f9364af1 on February 19, 2025.

Robb, M. B., & Mann, S. (2025). *Talk, trust, and trade-offs: How and why teens use AI companions.* Common Sense Media. Accessed at www.commonsensemedia.org/sites/default/files/research/report/talk-trust-and-trade-offs_2025_web.pdf on September 24, 2025.

Robinson, B., & Schaible, R. M. (1995). Collaborative teaching: Reaping the benefits. *College Teaching, 43*(2), 57–59.

Rock, M. L., Gregg, M., Thead, B. K., Acker, S. E., Gable, R. A., & Zigmond, N. P. (2009). Can you hear me now? Evaluation of an online wireless technology to provide real-time feedback to special education teachers-in-training. *Teacher Education and Special Education, 32*(1), 64–82. https://doi.org/10.1177/0888406408330872

Rock, M. L., Schumacker, R. E., Gregg, M., Howard, P. W., Gable, R. A., & Zigmond, N. (2014). How are they now? Longer term effects of eCoaching through online bug-in-ear technology. *Teacher Education and Special Education, 37*(2), 161–181. https://doi.org/10.1177/0888406414525048

Rogers, E. M., Singhal, A., & Quinlan, M. M. (2019). Diffusion of innovations. In D. W. Stacks, M. B. Salwen, & K. C. Eichhorn (Eds.), *An integrated approach to communication theory and research* (3rd ed., pp. 415–434). Routledge.

Romrell, D., Kidder, L. C., & Wood, E. (2014). The SAMR Model as a framework for evaluating mLearning. *Journal of Asynchronous Learning Networks, 18*(2).

Roose, K. (2024, October 24). Can A.I. be blamed for a teen's suicide? *The New York Times.* Accessed at www.nytimes.com/2024/10/23/technology/characterai-lawsuit-teen-suicide.html on April 11, 2025.

Rose, T. (2015). *The end of average: How we succeed in a world that values sameness.* HarperOne.

Rosenberg, N., & Huntington, R. N. (2021). Distance bug-in-ear coaching: A guide for practitioners. *Behavior Analysis in Practice, 14*(2), 523–533.

Rostand, E. (1898). *Cyrano de Bergerac* (G. Hall, Trans.). Doubleday.

Rumelhart, D. E., Hinton, G. E., & Williams, R. J. (1986). Learning representations by back-propagating errors. *Nature, 323*, 533–536. https://doi.org/10.1038/323533a0

Ruwe, T., & Mayweg-Paus, E. (2024). Embracing LLM feedback: The role of feedback providers and provider information for feedback effectiveness. *Frontiers in Education, 9*, Article 1461362. https://doi.org/10.3389/feduc.2024.1461362

Salessi, S., Omar, A., & Vaamonde, J. D. (2017). Conceptual considerations of work passion. *Ciencias Psicológicas, 11*(2), 165–178.

Salewski, L., Alaniz, S., Rio-Torto, I., Schulz, E., & Akata, Z. (2023). In-context impersonation reveals large language models' strengths and biases. In A. Oh, T. Naumann, & A. Globerson (Eds.), *Proceedings of the 37th International Conference on Neural Information Processing Systems* (pp.72044–72057). Curran Associates.

Sarason, S. B. (1972). *The creation of settings and the future societies*. Brookline Books.

Sato, W., Shimokawa, K., & Minato, T. (2025). Exploration of Mehrabian's communication model with an android. *Scientific Reports, 15*, Article 25986.

Schaaff, K., & Heidelmann, M.-A. (2024). *Impacts of anthropomorphizing large language models in learning environments*. arXiv. Accessed at https://arxiv.org/html/2408.03945v1 on April 11, 2025.

Scheeler, M. C., McKinnon, K., & Stout, J. (2012). Effects of immediate feedback delivered via webcam and bug-in-ear technology on preservice teacher performance. *Teacher Education and Special Education, 35*(1), 77–90.

Schindler, L. A., Burkholder, G. J., Morad, O. A., & Marsh, C. (2017). Computer-based technology and student engagement: A critical review of the literature. *International Journal of Educational Technology in Higher Education, 14*(1), Article 25.

Schiro, M. S. (2013). *Curriculum theory: Conflicting visions and enduring concerns* (2nd ed.). SAGE.

Schmidt, W. H., & Houang, R. T. (2012). Curricular coherence and the Common Core State Standards for mathematics. *Educational Researcher, 41*(8), 294–308.

Schmoker, M. (2023, February). We need coherent, teacher-built curriculum—NOW! *Educational Leadership, 80*(5).

Scott, T. (1986). *Top gun* [Film]. Don Simpson/Jerry Bruckheimer Films.

Scruggs, T. E., Mastropieri, M. A., & McDuffie, K. A. (2007). Co-teaching in inclusive classrooms: A metasynthesis of qualitative research. *Exceptional Children, 73*(4), 392–416.

Seo, S.-M. (2018). The effect of fine motor skills on handwriting legibility in preschool age children. *Journal of Physical Therapy Science, 30*(2), 324–327.

Shahid, S., & Din, M. (2021). Fostering psychological safety in teachers: The role of school leadership, team effectiveness & organizational culture. *International Journal of Educational Leadership and Management, 9*(2), 122–149.

Shams, L., & Seitz, A. R. (2008). Benefits of multisensory learning. *Trends in Cognitive Sciences, 12*(11), 411–417.

Shannon, C. E. (1950). Programming a computer for playing chess. *Philosophical Magazine, 41*(314), 256–275.

Sharma, S. V. (2006). High school students interpreting tables and graphs: Implications for research. *International Journal of Science and Mathematics Education, 4*, 241–268.

Shneiderman, B. (2022). *Human-centered AI*. Oxford University Press.

Shulman, L. S. (1986). Those who understand: Knowledge growth in teaching. *Educational Researcher, 15*(2), 4–14.

Shute, V. J. (2008). Focus on formative feedback. *Review of Educational Research, 78*(1), 153–189.

Shute, V. J., & Psotka, J. (1996). Intelligent tutoring systems: Past, present, and future. In D. Jonassen (Ed.), *Handbook of research on educational communications and technology* (pp. 570–600). Macmillan.

Simpson, E. J. (1966). *The classification of educational objectives, psychomotor domain* [Progress report on grant-supported project]. Accessed at https://files.eric.ed.gov/fulltext/ED010368.pdf on April 11, 2025.

Sizer, T. R., & Sizer, N. F. (1999). *The students are watching: Schools and the moral contract.* Beacon Press.

Skaalvik, E. M., & Skaalvik, S. (2007). Dimensions of teacher self-efficacy and relations with strain factors, perceived collective teacher efficacy, and teacher burnout. *Journal of educational psychology, 99*(3), 611.

Skinner, B. F. (2003). *The technology of teaching.* Skinner Foundation.

Sleeman, D., & Brown, J. S. (Eds.). (1982). *Intelligent tutoring systems.* Academic Press.

Smakman, M. H. J., Konijn, E. A., & Vogt, P. A. (2022). Do robotic tutors compromise the social-emotional development of children? *Frontiers in Robotics and AI, 9,* Article 734955.

Sparrow, B., Liu, J., & Wegner, D. M. (2011). Google effects on memory: Cognitive consequences of having information at our fingertips. *Science, 333*(6043), 776–778.

Stanford Teaching Commons. (n.d.). *Understanding AI literacy.* Accessed at https://teachingcommons.stanford.edu/teaching-guides/artificial-intelligence-teaching-guide/understanding-ai-literacy on August 9, 2025.

Staudt Willet, K. B., & He, D. (2024). Educators' invisible labour: A systematic review. *Review of Education, 12*(2), Article e3473.

Steinbeck, J. (1962). *Travels with Charley: In search of America.* Penguin Books.

Stigler, J. W., & Hiebert, J. (2009). *The teaching gap: Best ideas from the world's teachers for improving education in the classroom.* The Free Press.

Stravinsky, I. (1947). *Poetics of music: In the form of six lessons* (A. Knodel & I. Dahl, Trans.). Harvard University Press.

Sweller, J. (1988). Cognitive load during problem solving: Effects on learning. *Cognitive Science, 12*(2), 257–285.

Sweller, J., van Merriënboer, J. J. G., & Paas, F. (2019). Cognitive architecture and instructional design: 20 years later. *Educational Psychology Review, 31*(2), 261–292.

Tabibian, B., Upadhyay, U., De, A., Zarezade, A., Schölkopf, B., & Gomez-Rodriguez, M. (2019). Enhancing human learning via spaced repetition optimization. *PNAS, 116*(10), 3988–3993.

Teachers Pay Teachers. (n.d.). *Pencil borrowing.* Accessed at www.teacherspayteachers.com/browse?search=pencil%20borrowing on August 5, 2025.

Thorndike, E. L. (1920). A constant error in psychological ratings. *Journal of Applied Psychology, 4*(1), 25–29.

Tillmanns, T., Salomão Filho, A., Rudra, S., Weber, P., Dawitz, J., Wiersma, E., et al. (2025). Mapping tomorrow's teaching and learning spaces: A systematic review on genAI in higher education. *Trends in Higher Education, 4*(1), Article 2.

Toppo, G. (2024, August 7). *AI 'companions' are patient, funny, upbeat—and probably rewiring kids' brains.* The 74. Accessed at www.the74million.org/article/ai-companions-are-patient-funny-upbeat-and-probably-rewiring-kids-brains on April 11, 2025.

Toppo, G. (2025, July 24). *'Cognitive science,' all the rage in British schools, fails to register in U.S.* The 74. Accessed at www.the74million.org/article/cognitive-science-all-the-rage-in-british-schools-fails-to-register-in-u-s on August 20, 2025.

Unisys. (2024). *The AI equation: 2024 AI business impact research.* Author. Accessed at www.unisys.com/siteassets/collateral/report/2024-ai-insight-report.pdf on April 11, 2025.

Usma-Wilches, J. A. (2006). *Teacher autonomy: A review of the research literature.* University of Northern Iowa Graduate Research Papers. 1634.

Vallerand, R. J., & Reid, G. (1984). On the causal effects of perceived competence on intrinsic motivation: A test of cognitive evaluation theory. *Journal of Sport Psychology, 6*(1), 94–102.

VanLehn, K. (2006). The behavior of tutoring systems. *International Journal of Artificial Intelligence in Education, 16*(3), 227–265.

VanLehn, K. (2011). The relative effectiveness of human tutoring, intelligent tutoring systems, and other tutoring systems. *Educational Psychologist, 46*(4), 197–221.

Vembye, M. H., Weiss, F., & Bhat, B. H. (2024). The effects of co-teaching and related collaborative models of instruction on student achievement: A systematic review and meta-analysis. *Review of Educational Research, 94*(3), 376–422.

Venter, J., Coetzee, S. A., & Schmulian, A. (2025). Exploring the use of artificial intelligence (AI) in the delivery of effective feedback. *Assessment and Evaluation in Higher Education, 50*(4), 516–536.

Vidrin, I. (in press). Tender robotics: A framework for meaning-making through emotional investment. In E. Mullis, S. Skybetter, H. Bergen, & B. Simon (Eds.), *Robot theatre.* Routledge.

Vygotsky, L. S. (1978). *Mind in society: The development of higher psychological processes.* Harvard University Press.

Walker, D. M. (2022). *Reconsidering professional development and its impact on teacher learning: An examination of teacher motivation in a self-directed model of teacher professional development.* Montclair State University.

Walther-Thomas, C. S. (1997). Co-teaching experiences: The benefits and problems that teachers and principals report over time. *Journal of Learning Disabilities, 30*(4), 395–407.

Wang, J., & Fan, W. (2025). The effect of ChatGPT on students' learning performance, learning perception, and higher-order thinking: Insights from a meta-analysis. *Humanities and Social Sciences Communications, 12,* Article 621.

Wang, R. E., Ribeiro, A. T., Robinson, C. D., Loeb, S., & Demszky, D. (2024). *Tutor CoPilot: A human-AI approach for scaling real-time expertise.* arXiv. Accessed at https://arxiv.org/pdf/2410.03017 on April 11, 2025.

Warner, J. (2025). *More than words: How to think about writing in the age of AI.* Basic Books.

Warren, L. (2000). Teacher planning: A literature review. *Educational Research Quarterly, 24*(2), 37–42.

Wason, P. C. (1960). On the failure to eliminate hypotheses in a conceptual task. *Quarterly Journal of Experimental Psychology, 12*(3), 129–140.

Watkins, M. D. (2012, June). How managers become leaders. *Harvard Business Review, 90*(6), 64–72, 144.

Watters, A. (2021). *Teaching machines: The history of personalized learning.* The MIT Press.

Watzlawick, P., Bavelas, J. B., & Jackson, D. D. (1967). *Pragmatics of human communication: A study of interactional patterns, pathologies, and paradoxes.* Norton.

Wayback Machine. (n.d.). *The reusability paradox.* Accessed at https://web.archive.org/web/20050620084314/http://rclt.usu.edu/whitepapers/paradox.html on April 14, 2025.

Weber-Wulff, D., Anohina-Naumeca, A., Bjelobaba, S., Foltýnek, T., Guerrero-Dib, J., Popoola, O., et al. (2023). Testing of detection tools for AI-generated text. *International Journal for Educational Integrity, 19,* Article 26.

Weerts, M. (2020). *Co-teaching: Benefits and challenges of co-teaching in middle school* [Doctoral dissertation, St. Cloud State University]. The Repository at St. Cloud State. https://repository.stcloudstate.edu/cgi/viewcontent.cgi?article=1085&context=edad_etds

Westerman, D. A. (1991). Expert and novice teacher decision making. *Journal of Teacher Education, 42*(4), 292–305.

Wiggins, G. (1993). *Assessing student performance: Exploring the purpose and limits of testing.* Jossey-Bass.

Wiggins, G. (2012, September). Seven keys to effective feedback. *Educational Leadership, 70*(1), 10–16.

Wiggins, G., & McTighe, J. (2005). *Understanding by design* (Expanded 2nd ed.). ASCD.

Wiliam, D. (2012, September 1). *Feedback: Part of a system.* Accessed at www.ascd.org/el/articles/feedback-part-of-a-system on April 14, 2025.

Williamson, M. R., Fries, R., & Zhou, H. (2016). Long-term effectiveness of radar speed display signs in a university environment. *Journal of Transportation Technologies, 6*(3), 99–105.

Willingham, D. T. (2009). *Why don't students like school? A cognitive scientist answers questions about how the mind works and what it means for the classroom.* Jossey-Bass.

Winslow, B. (2011). *Effects of a structured handwriting readiness program on the fine motor skills of children in Head Start* [Master's thesis, East Carolina University]. The ScholarShip. Accessed at https://thescholarship.ecu.edu/items/5438f698-e3d9-4ecc-bcec-25cffdf48d2b on November 11, 2025.

Wisniewski, B., Zierer, K., & Hattie, J. (2020). The power of feedback revisited: A meta-analysis of educational feedback research. *Frontiers in Psychology, 10,* Article 3087.

Workforce Lab. (2024, June). *The workforce index.* Author. Accessed at https://d34u8crftukxnk.cloudfront.net/slackpress/prod/sites/6/June-2024-Workforce-Index-from-the-Workforce-Lab-at-Slack.pdf on April 11, 2025.

Worth, J., & Van den Brande, J. (2020, January). *Teacher autonomy: How does it relate to job satisfaction and retention?* National Foundation for Educational Research. Accessed at https://files.eric.ed.gov/fulltext/ED604418.pdf on September 19, 2025.

Wu, X., Saraf, P. P., Lee, G., Latif, E., Liu, N., & Zhai, X. (2025). Unveiling scoring processes: Dissecting the differences between LLMs and human graders in automatic scoring. *Technology, Knowledge and Learning.* Advance online publication.

Yoon, S.-Y., Miszoglad, E., & Pierce, L. R. (2023). *Evaluation of ChatGPT feedback on ELL writers' coherence and cohesion.* arXiv. Accessed at https://arxiv.org/abs/2310.06505 on April 14, 2025.

Zee, M., & Koomen, H. M. Y. (2016). Teacher self-efficacy and its effects on classroom processes, student academic adjustment, and teacher well-being: A synthesis of 40 years of research. *Review of Educational Research, 86*(4), 981–1015.

Zelazo, P. D., Blair, C. B., & Willoughby, M. T. (2016). *Executive function: Implications for education* (NCER 2017-2000). National Center for Education Research. Accessed at https://ies.ed.gov/ncer/2025/01/executive-function-implications-education on October 1, 2025.

Zhang, N., & Leong, W. Y. (2025). Intelligent emotional computing with deep convolutional neural networks: Multimodal feature analysis and application in smart learning environments. *EURASIA Journal of Mathematics, Science and Technology Education, 21*(8), Article em2680.

Zheng, J., & Meister, M. (2025). The unbearable slowness of being: Why do we live at 10 bits/s? *Neuron, 113*(2), 192–204.

Zöller, N., Berger, J., Lin, I., Fu, N., Komarneni, J., Barabucci, G., et al. (2025). Human-AI collectives most accurately diagnose clinical vignettes. *PNAS, 122*(24), Article e2426153122.

Index

C

D

E

F

U

W